Contemporary Urban Design Thinking

This series is indexed in SCOPUS

This series investigates contemporary insights in urban design theory and practice. Urbanism has considerably changed and developed over the years and undergoes a transformation moving into an era of uncertainty, regeneration, and resilience. In the recent past urban planning and design has focused on growth and was economically driven. This is no longer feasible in the context of rapid change, environmental and social transformation, and climate impacts. Therefore, new ideas for creating innovative approaches and implementation in urbanism are urgently required.

This series publishes titles dealing with novel methods of urbanism such as nature-driven urbanism, social-driven urban planning, and landscape-based urban design.

The series includes books and contributions by established researchers and leaders in the fields of urban design, city development and landscape urbanism. The books contain the most recent insights into urbanism and provides actual and timely publications that fill the gap in current literature.

The series is of special interest to urbanists, landscape architects, architects, policy makers, city/urban planners, urban designers/researchers, and to all of those interested in a wide-ranging overview of contemporary urban design innovations in the field.

Rob Roggema
Editor

Dry Urbanism

Designing for Drought in the City

Springer

Editor
Rob Roggema
Escuela de Arquitectura, Artes y Diseño
Tecnológico de Monterrey
Monterrey, Nuevo León, Mexico

ISSN 2522-8404 ISSN 2522-8412 (electronic)
Contemporary Urban Design Thinking
ISBN 978-3-031-77346-4 ISBN 978-3-031-77347-1 (eBook)
https://doi.org/10.1007/978-3-031-77347-1

This Springer imprint is published by the registered company Springer Nature Switzerland AG
The registered company address is: Gewerbestrasse 11, 6330 Cham, Switzerland

Contents

Editor and Contributors

About the Editor

Rob Roggema is a Distinguished Professor of Regenerative Culture at Tecnológico de Monterrey, Monterrey, Mexico, and the Principal Consultant and Director/Founder of Cittaideale, office for adaptive design and planning. He is a Visiting Professor at Queen's University Belfast and the lead author of the "Architecture, Urban Design and Planning" chapter of the third assessment report of the UCCRN (Urban Climate Change Research Network). He is a Landscape Architect and an internationally renowned design expert on regenerative design, sustainable urbanism, climate adaptation, energy landscapes and urban agriculture. He has previously held several professorship positions at universities in the Netherlands and Australia, State and Municipal governments and design consultancies. Rob developed the Swarm Planning concept, a dynamic way of planning the city for future adaptation to climate change impacts.

Rob was the leader of the climate adaptive design 2021 team for the Groningen region ('Moeder Zernike'), initiated the Food-Energy-Water-nexus project 'the Moveable Nexus' (SUGI/JPI-Europe) and led the design team of 'Nature-Rich Netherlands' and Greening NEOM. For his PhD, he developed the Swarm Planning concept, a dynamic way of planning urban landscapes for future adaptation to climate change impacts.

Rob has designed and led over 30 design charrettes around the world, involving communities, academics, governments and industries in design processes for more resilient communities. He is the series editor of the Contemporary Urban Design Thinking series. Additionally, he has written books on nature- and landscape-driven planning and design of landscapes and cities, climate adaptation and design, Urban Agriculture, and about design charrettes, Rio's FoodRoofs and Design for Recovery in Japan.

Contributors

Kristel Aalbers, M.Sc. (section of Environmental Technology and Design), is an Urbanist and Architect specializing in sustainable urban planning and design with a focus on water. In her prior jobs as senior advisor of urban planning at the water board of Delfland and the municipality of Delft, she concerned herself with the (spatial) integration of (surface) water, ecology and slow traffic in urban plans for newly built developments and mainly in-town renewals.

As coordinator of the master track Urbanism at the Faculty of Architecture and the Built Environment at the TU Delft, she is responsible for the Urbanism two-year MSc curriculum. As a lecturer in the Urbanism track, she mentors students in R&D projects.

Prof. Ali A. Alraouf is an architect and urban designer focusing on research and consultancy within the domain of holistic sustainability in architecture and urbanism. He was a Visiting Scholar at the University of California at Berkeley, USA. Alraouf has held permanent and visiting teaching and research positions at regional and international universities. Alraouf's current research interests are knowledge-based urban development, post-carbon contemporary Gulf cities, sustainable urban vitalization of historical districts and knowledge and creative cities. He provides consultancy in numerous projects regionally and internationally and act as a consultant for international organizations including UN-Habitat, UNESCOW, ESCOWA and ICOMOS among others. He published more than 150 journal refereed papers, critical reviews and essays, in addition to books and book chapters. Alraouf serves in editorial boards and acts as a reviewer for international journals and conferences. He has been invited to present his research work at international institutions in over 30 countries. Alraouf is the recipient of several research and teaching awards. In addition to teaching at HBK University in Qatar, Alraouf currently acts as senior advisor for research and development at the Ministry of Municipality in Qatar. He is also a board member of the International Society for City Planners (ISOCARP).

Dr. Robert D. Bornstein is an Emeritus Professor in the Department of Meteorology at San Jose State University. He is a leader and influential researcher in the field of urban meteorology and climatology, focusing on observation analysis and modeling of urban boundary layers. He is considered one of the pioneers and most influential researchers in urban weather and climatology. Dr. Bornstein has published more than 200 peer-reviewed papers, book chapters and other writings and has received numerous research grants for his research.

Catalina Cavazos is an architect with over eight years of experience in landscape architecture and urban design. As the Manager of PEW STUDIO, she coordinates multidisciplinary teams for complex projects across Mexico, promoting flexible and equitable public space design. Her work includes impactful projects like Plaza Los Compositores and Parque Brisas. She has also led strategic and creative efforts for residential master plans with clients like COCONSA and Ruba. Catalina frequently speaks at universities and organizations on sustainable urban design and gender perspectives.

Rosa de Wolf Ir. Rosa de Wolf is a Ph.D. candidate at Delft University of Technology's Urban Ecology Design Lab. Her research focuses on the potential of multi-functionality within the transformative design of Dutch business districts into Workscapes. Besides her work, she volunteers for Africa Wood Grow to contribute to reforestation in Kenya by empowering local communities. Rosa graduated cum laude from the Urbanism Master track at Delft University of Technology with her project Urban Arid Green.

Montserrat Flores is an architect who graduated from Tecnológico de Monterrey with over three years of experience in landscape architecture and urban design. As Project Coordinator at PEW STUDIO, she advocates for flexible and equitable public space planning. Her work includes notable projects like Plaza Los Compositores and Parque Brisas, and she has collaborated with clients like COCONSA and Ruba. Montserrat is also active in community service, contributing to high-impact projects through Rotary International's Rotaract Del Valle Monterrey.

Diana Garcia-Cejudo, M.Sc. Urb, M.Arch. is an Assistant Professor and Consultant at Tecnologico de Monterrey Campus Querétaro. She studied architecture at Tec de Monterrey Querétaro (México) and has a post-graduate Master's in Urbanism from the European Masters in Urbanism (EMU) from TU Delft and a Master's in Architecture from Tec de Monterrey Querétaro, México. Her current research focuses on water-wise cities, water-sensitive design and climate resilience in the global south.

She coordinates with Rodrigo Pantoja, one of the graduation studios related to Water-Wise Cities and Urban Resilience through spatial applications of green and blue infrastructure and nature-based solutions.

For the past two years, she has been a Senator representing the School of Architecture, Art, and Design in the institution's Academic Senate.

Additionally, she is a Citizen Advisor in the Municipal Council of Urbanism and Mobility in Querétaro and an active member of the environmental committee at the Querétaro River Council.

She is the Founder and Director of ABC Investeam, a real estate company whose work focuses on consultancy on real estate projects and urban development.

Through the tactical urbanism collective, she has directed temporary collective interventions to regenerate abandoned public spaces in the city.

Jorge Gerini Igartua is an architect and urban designer with over 15 years of experience working on projects at diverse scales and in various contexts worldwide. His expertise spans sustainable urbanism, master planning and resilient urban environments. He has worked internationally on projects such as self-sufficient housing and zero-carbon mixed-use developments in the UK and the Netherlands, as well as large-scale masterplans in the USA, Brazil, Paraguay and the Middle East. Jorge's work also extends to leisure-led developments in Mexico, the Dominican Republic, Belize and the Bahamas.

Currently, Jorge serves as Urban Design Director at Sordo Madaleno Architects, where he oversees key projects, including the San Antonio and Paseo Polanco masterplans in Mexico City. In addition to his design leadership, Jorge plays a key role in aligning project goals with broader development strategies, ensuring that both client objectives and the firm's ambitions are integrated from the outset.

Jorge holds a Bachelor of Architecture from Universidad de Guadalajara (UDG) and a Master's in Sustainable Urbanism from the Bartlett School of Architecture, University College London (UCL). His academic and professional interests include sustainable urban design, the integration of ecological principles in urban planning and the creation of livable urban spaces in diverse global contexts.

Dr. Sue C. Grady received her Ph.D. in Earth and Environmental Sciences from the City University of New York. Her Ph.D. included training in health and medical geography at Hunter College. Dr. Grady has an M.P.H. degree in International Health from Tulane University and an M.A. degree in Medical Anthropology from Hunter College. She earned a B.A. degree in Medical Geography from the University of Minnesota, where she began her career as a registered nurse in Minneapolis. Dr. Grady conceptualizes health and disease from a human ecological perspective to study place-based clinical (individual) and public health (population) outcomes.

Dr. Ibrahim Hegazy is a Professor of Urban Planning and Design at King Abdulaziz University. He has a strong academic background and extensive experience in development and environmental management. His expertise, developed through projects and training in the UK, focuses on regional planning and environmental assessment. Dr. Hegazy contributes to participatory strategic planning for urban, regional and national environmental projects and educates on urban design methodologies. He has been a faculty member at Mansoura University (Egypt) since 2001 and has published over 70 papers in international journals and conferences on sustainable development and environmental planning.

Dr. Solmaz Hosseinioon is an Iranian urban designer, planner and resilience expert. She has over two decades of extensive experience in multidisciplinary development aspects across multiple scales, from national to local levels. Her expertise spans urban plans and policies, resilience, sustainable development, heritage, placemaking, informality and disaster management. She holds a Ph.D. from the University of Melbourne and dual Master's degrees in urban design and architecture from Iran. She has made significant contributions to academia and professional education in Iran. In addition to her academic publications, she has prepared numerous strategic documents and reports for industry and public national and international institutions. Having come from the ancient country of Iran, she is passionate about reviving the knowledge and wisdom of the past for future generations towards sustainable and resilient living environments.

Dr. Hossam Ibrahim is a distinguished Urban and Regional Planning Advisor with over 26 years of experience in the field. He has played a pivotal role in shaping urban and regional landscapes across Egypt, the UK and Qatar, working with government sectors, private consultancies and academia. Currently serving as the Senior Urban and Regional Planning Advisor and Head of the Municipalities Spatial Development Plans Unit at the Ministry of Municipality in Qatar, Dr. Hossam has significantly contributed to Qatar's planning legislation, housing strategies and coastal management frameworks.

His expertise spans various domains, including urban development, industrial strategies and tourism planning. Dr. Hossam is also a published academic with over 40 refereed journal papers and book chapters. His leadership and dedication to sustainable urban development are reflected in his role as a principal advisor on key national projects, such as the Qatar National Housing Strategy and the Qatar Urban Planning Law. Dr. Hossam's commitment to knowledge sharing and capacity building underscores his influence on the future of urban planning in the MENA region.

Rodrigo Junco is a dedicated urban planner with a passion for creating more livable, safe and sustainable cities. As a graduate of Urbanism from the Tecnológico de Monterrey, Rodrigo has demonstrated his commitment to urban development through a diverse array of professional and academic experiences, showcasing his ability to address complex urban challenges with innovative and impactful solutions.

As a Project Coordinator at the Faculty of Excellence at Tecnológico de Monterrey, he plays a key role in leading projects such as the Green Metropolis. His internships at IBI Group Mexico and Distrito TEC further deepened his understanding of urban design and analysis across different scales, enhancing his expertise in the field.

Rodrigo has been recognized with prestigious awards, including the "Acceleration Award" and "Best Collaborators Award" at the Next Generation City Action of the UIA International Congress of Architecture in Copenhagen 2023. He was also a finalist in the First State Competition Against Climate Change in Nuevo León. His active engagement in professional communities, such as the Architectural Editorial Committee of Cemex México and the American Planning Association, underscores his commitment to advancing sustainability and innovation in urban environments.

Dr. Jun Hyun Kim is Director and Professor in the School of Planning, Design and Construction (SPDC) at Michigan State University (MSU). Previously, he served as Program Director of both the Landscape Architecture and the Master of Environmental Design programs at MSU. Prior to joining MSU through a targeted search, he held positions as an Assistant Professor at Texas A&M University (TAMU) and California Polytechnic State University at San Luis Obispo. Dr. Kim has received numerous awards for his outstanding contributions to the field, including the Excellence in Design Studio Teaching Award (senior level) from the Council of Educators in Landscape Architecture (CELA). Dr. Kim's research focuses on the impacts of environments on human and ecological health, enhancing community resilience

Dr. Nathan Moore completed his B.A. in Physics in 1993 from the University of Virginia and received an M.S. in Physics in 1997 from the University of Oregon. He earned his Ph.D. in Earth and Ocean Sciences from Duke University in 2004. Dr. Moore has ongoing research projects in China, East Africa and the Amazon. His research interests include regional climate modeling, Land Cover/Land Use Change, integrating models of human and natural systems, food security, remote sensing and water issues.

Miguel Ortiz is a student of Urbanism at Tecnológico de Monterrey with a strong interest in research, regenerative urban design and the impact of public policies on urban development. He is currently working as an environmental consultant on LEED construction projects. Over the past few years, he has been involved in two significant research projects: one focused on understanding the perceptions, motivations and incentives of stakeholders involved in the rehabilitation of creeks and the other aimed at analyzing how loneliness affects urban quality of life and the perception of green spaces. In addition, he has taken part in international events with schools from China and Canada. As an urbanist, Miguel is dedicated to using creativity and empathy as fundamental tools to inspire and guide his projects, always with the vision of improving the urban environment for the well-being of all.

Rodrigo Pantoja-Calderon, MLA Rodrigo studied Architecture, Art History and Visual Arts at Drury University in Springfield, Missouri and Volos, Greece. He also took a masterclass in landscape urbanism at the Berlage Institute in Rotterdam. Rodrigo holds a Master's degree in Landscape Architect from the Universidad Iberoamericana.

He collaborated with Shigeru Ban Architects in Japan and Wiel Arets Architects in the Netherlands. He has written articles on architecture for various media such as La Jornada, Provincia, el Financiero, Plaza de Armas, Diario de Querétaro and Obras.

Rodrigo has been a fellow for the National Fund for Culture and the Arts (FONCA) and the Program for the Encouragement of Artistic Creation and Development (PECDA) on four occasions, developing projects that promote the recovery of green areas, urban gardens and public spaces. His work as an architect, consultant and landscaper has been for the private and public sectors, collaborating with various municipalities, state governments and organizations such as the German Cooperation Agency (GIZ) and Infonavit.

Rodrigo's projects promote environmental and social revitalization through sustainable proposals and Codesign. His architectural methodology includes understanding normative, social, environmental and economic components through participatory design, statistical analysis and public space intervention.

He is the Founder and Director of the architecture firm EVO-A-LAB and the Academic Program Director of Architecture at the Tecnologico de Monterrey in Querétaro (2011-2016 and 2018).

Brenda Perez is the Director of Pew Studio and holds a Master's degree in City Planning with a specialization in Green Infrastructure from the University of Melbourne. Throughout her career, she has championed urban landscape design as a guiding thread in broader urban design strategies.

She worked on urban transport and sustainable energy projects, and since founding PEW Studio in 2014, she has led award-winning projects, including public spaces and residential master plans across Mexico.

Dr. Ana Rivera is an ORISE Postdoctoral Researcher at the US Environmental Protection Agency's Center for Public Health and Environmental Assessment. She earned her Ph.D. in Geography from Michigan State University, an M.A. in Geography from San Jose State University and an M.S. in Environmental Systems from Tecnologico de Monterrey. Dr. Rivera has been a consultant for the World Bank's program on Health, Climate, Environment, and Disasters, as well as for the RAND Corporation. She has also been a member of NASA's Health and Air Quality Applied Sciences Team (HAQAST). Ana uses geospatial technology to analyze climate-social interaction in urban settings, focusing on extreme heat weather events and their relation to demographic, socioeconomic and institutional issues. She is currently pursuing a Master's in Strategic Foresight at Tecnologico de Monterrey's School of Government and Public Transformation.

Lizbeth Anahi Saavedra Flores is a Mexican architect and urban designer specializing in sustainable urban development. She holds a Master of Architecture (M.Arch.) in Sustainable Urban Design from the University of Nottingham and a Bachelor of Architecture (B.Arch.) from Universidad Autónoma de México.

Her work focuses on integrating ecological and social considerations into urban design, with a particular emphasis on urban retrofit, regeneration and landscape-driven master planning at both local and territorial scales. This work earned her recognition at the XVI Mexico Architecture Biennale. Her research emphasizes water sensitivity and urban resilience, particularly in the restoration of Mexico City's lacustrine landscapes.

Since joining Sordo Madaleno in 2019, she has contributed to major urban projects and fostered interdisciplinary collaboration between research and practice. By building strategic partnerships with architectural, cultural and academic institutions, she bridges theory and design, advancing sustainable urbanism across diverse contexts.

Esteban Sheridan has extensive experience in journalism, television, advertising and marketing. He currently serves as the manager for RRG Nature Based Solutions' Mexico office, a sustainability firm specializing in the decarbonization of global supply chains through nature-based solutions, innovative business practices and long-term offtake contracts. RRG NBS manages 60,000 hectares of sustainably farmed land throughout the Americas and oversees 100,000 hectares of conservation and restoration projects. The firm is advancing five carbon offset projects towards full certification and produces 826 million cubic meters of sustainable water supplies annually.

Nico Tillie Dr. ir. Nico Tillie (Ph.D.) is a Landscape Architect and plant specialist and specializes in synergetic urban landscape planning at TUDelft, the Netherlands, where he leads the Urban Ecology Design Lab. He works on issues ranging from low carbon cities, climate adaptation and urban agriculture to biodiversity. For many years, he worked for the city of Rotterdam in urban green and sustainability-related projects.

Apart from being a Senior Fellow of the Global Cities Institute of the University of Toronto in Canada, he was in the advisory board of the sustainable development of the Olympic Games Paris 2024. In 2022, he was the conference chair of the Ecocities World Summit.

Wendy Walls is a landscape architect researcher, writer and educator. She holds a Ph.D. in climate design and is a lecturer at the University of Melbourne. Her research focuses on landscape design methods under threat of climate change and includes the role of landscape architecture in designing for the lived experience of heating cities. She has explored this through developing data-driven and digital design methodologies informed by eco-critical theory and material explorations. She pursues this work through design teaching, alongside collaboration with design practice and community. Outcomes from her research have been published and presented at peer-reviewed international events including the *Journal of Landscape Architecture*, *Digital Landscape Architecture*, IFLA and the Architectural Science Association Conference.

Chapter 1
Designing Arid Cities: Cope, Embrace, Reverse

Rob Roggema

Abstract In recent decades, climate change has caused an increase in dry landscapes. These landscapes are reaching and entering urban environments, causing heat islands, drylands, and unhealthy environment for the urban inhabitants. The acceptance of this ongoing trend will lead to a suffering urban population and loss of life. The city has, so far, not yet been seen as the playground for transforming arid cityscapes into attractive and livable places. Three strategies have been identified to deal with and act on the increase in dry urban conditions. First, cope with the circumstances, e.g., create safe havens for urban inhabitants to survive the heat and drought. Second, embrace the dry conditions by welcoming and integrating the drought in species, xeriscapes, and culture in the urban context. And finally, reverse the dry conditions by implementing forests and wetlands that can reduce the temperatures and provide humidity. These three strategies shall be used in combination and in parallel.

Keywords Arid · Drylands · Urban design · Planning · Landscape · Xeriscape · Desertification

1.1 Introduction

Climate change causes an increase in average temperatures. According to the IPCC, this process has been going on for a long time, at least since 1860, and is accelerating in recent decades (IPCC 2023). It is expected that the temperatures will rise further, with a chance of more than 50% average temperatures on earth to rise with more than 1.5 degree Celsius before 2040 (IPCC 2018). In many places, this leads to hotter and dryer conditions, especially in cities (Yuan et al. 2022).

R. Roggema (✉)
Escuela de Arquitectura, Artes y Diseño, Tecnológico de Monterrey, Monterrey, Nuevo León, Mexico
e-mail: rob@cittaideale.eu

R. Roggema (ed.), *Dry Urbanism*, Contemporary Urban Design Thinking,
https://doi.org/10.1007/978-3-031-77347-1_1

In three ways cities are exposed to warming conditions. The general temperature rise hits cities as it does everywhere. Cities contain a surplus of sealed pavements, concrete, and asphalt which adds up to the creation of urban heat islands (UHIs). This results in exaggeration of temperature rise in concretized places and the city in general, and the occurrence of more heat island effects in cities worldwide (Mentaschi et al. 2022). Secondly, many cities are still expanding in area, which implies that a larger number of people and urban environments are entering these hotter conditions (Gao et al. 2024). Not only is it getting hotter in urban environments, also more people are therefore exposed to these conditions (Tuholske et al. 2021). The third way cities face arid conditions, is that more landscapes around cities are increasingly arid (Cherlet et al. 2018; Gaur and Squires 2018; Chen et al. 2023). The dry landscape, sometimes a desert is then reaching the cities' boundaries or even enters the urban area (Egidi et al. 2021).

About 20% of arid landscapes are degrading due to climate change and human actions (Burrell et al. 2020). The main factors accelerating these processes are (Cherlet et al. 2018):

- Extreme climate events, changes in precipitation and temperature, and seasonal variability. Where some areas become dryer other turn more humid (Huang et al. 2016; Prăvălie et al. 2019), moreover these are uncertain processes (Lian et al. 2021).
- Increasingly intense use of the land, such as grazing, water extraction, deforestation, agriculture, and urbanization.
- Local disruptions, such as fires, insect outbreaks and biotic invasions (Rillig et al. 2023).

Globally, the growth of arid areas increases depending on the future climate. When global warming will be 1.5 °C, arid areas replace 1.1% of other climate zones (eg. 1.6 million km^2). When temperatures rise further larger areas turn arid. At 2 °C this is 1.4% and 2.1 million km^2, at 3 °C 3% or 4.6 million km^2, and 3.8% at 4 °C, or 5.6 million km^2 (Spinoni et al. 2021).

Meanwhile, semi-arid areas have expanded also. Their growth between 1990 and 2004 relative to the period 1948–1962 has been 7%, or 0.8 million km^2 (Huang et al. 2016).

This ongoing increase of arid regions is visible on the global map (Fig. 1.1), with large areas in the southern US and Mexico, Southern South America, Northern and Southern Africa, the Middle East, large parts of Russia, Kazakhstan, India and China, and Australia.

The aridification of the world has a direct impact on water stress. Where dry landscapes meet places in which a large number of people live, water scarcity turns into stress. The most severe water stress (Fig. 1.2) occurs in 25 countries, hitting a quarter of the global population, often the most vulnerable ones (Kuzma et al. 2023). Many of the chapters in this book focus on these countries, such as Mexico, California, Morocco, the Middle East and Australia.

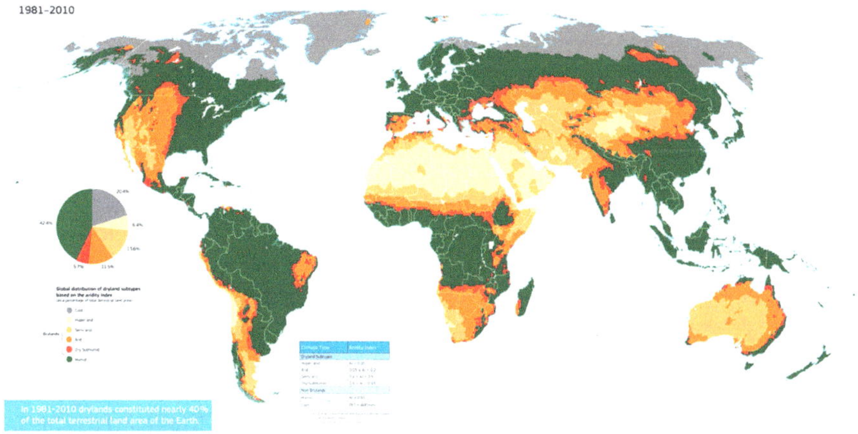

Fig. 1.1 Global growth and development of arid regions. (Source: WAD3-JRC, Cherlet et al. 2018)

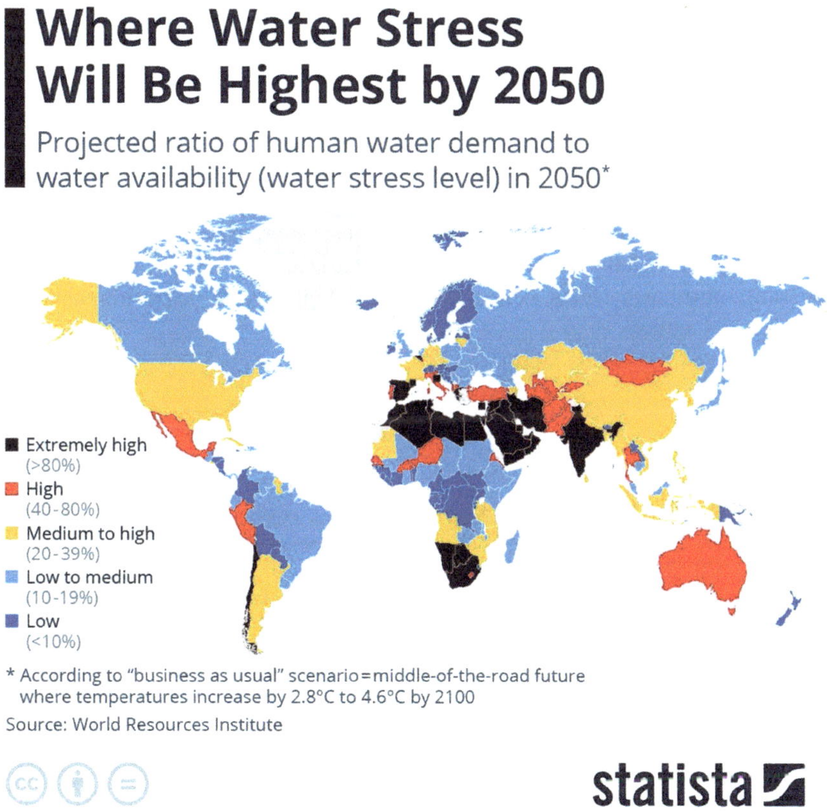

Fig. 1.2 Global water stress. (Source: Kuzma et al. 2023)

1.2 Urban Strategies

The main question is how to act and respond to these developments. How can we design cities, its neighborhoods, and public spaces that suffer from droughts, heat and water stress? This book presents different strategies on how we can approach this.

1. Coping with dry conditions. The first strategy is to cope. The dry and hot conditions in the city occur and are a nuisance. To be able to live through this, measures should be taken that compensate or relief the impact. This can be achieved by creating rain gardens, green parks with tree cover, amenities for shelter and shade or technologies for providing humidity (sprinklers, water dispersers, fountains, etc.).
2. Embrace dry conditions. This strategy aims to accept and embrace the arid circumstances. The city becomes as it were a part of the desert. This can be achieved by creating xeriscapes (arid cityscapes), integrating dry-loving species in public spaces, and using cultural understanding of the desert history to guide contemporary urban design and architecture.
3. Reverse dry conditions. The third strategy aims to reverse the arid climate. Instead of accepting harsh and dry circumstances, this strategy reverses this, and aims to create more humid and cooler conditions in and surrounding the city. This can be achieved by arranging and planting large scale climate forests, water systems, and minimizing impermeable surfaces.

Each of these strategies has their merits and should be used in combination while strengthening each other. In the city, an intelligent interplay between creating greener and humid public spaces, allowing the desert to manifest itself in the urban context and creating extensive green belts and intertwined ecological corridors around and through the city, must be enhanced to reduce the negative consequences of aridity in urban systems, hence providing the inhabitants an environment that is livable and healthy.

References

Burrell AL, Evans JP, De Kauwe MG (2020) Anthropogenic climate change has driven over 5 million km^2 of drylands towards desertification. Nat Commun 11:3853. https://doi.org/10.1038/s41467-020-17710-7

Chen Y, Lu H, Wu H et al (2023) Global desert variation under climatic impact during 1982–2020. Sci China Earth Sci 66:1062–1071. https://doi.org/10.1007/s11430-022-1052-1

Cherlet M, Hutchinson C, Reynolds J, Hill J, Sommer S, von Maltitz G (eds) (2018) World atlas of desertification. Publication Office of the European Union, Luxembourg

Egidi G, Quaranta G, Salvia R, Salvati L, Včeláková R, Cudlín P (2021) Urban sprawl and desertification risk: unraveling the latent nexus in a mediterranean country. J Environ Plan Manag 65(3):441–460. https://doi.org/10.1080/09640568.2021.1886913

Gao S, Chen Y, Chen D, He B, Gong A, Hou P, Li K, Cui Y (2024) Urbanization-induced warming amplifies population exposure to compound heatwaves but narrows exposure inequality between global north and south cities. NPJ Clim Atmos Sci 7(154). https://doi.org/10.1038/s41612-024-00708-z

Gaur M, Squires V (2018) Geographic extent and characteristics of the world's arid zones and their peoples. In: Gaur M, Squires V (eds) Climate variability impacts on land use and livelihoods in drylands. Springer, Cham. https://doi.org/10.1007/978-3-319-56681-8_1

Huang J, Ji M, Xie Y, Wang S, He Y, Ran J (2016) Global semi-arid climate change over last 60 years. Clim Dyn 46:1131–1150. https://doi.org/10.1007/s00382-015-2636-8

IPCC (2018) Summary for policymakers. In: Global warming of 1.5°C. An IPCC special report on the impacts of global warming of 1.5°C above pre-industrial levels and related global greenhouse gas emission pathways, in the context of strengthening the global response to the threat of climate change, sustainable development, and efforts to eradicate poverty [Masson-Delmotte V, Zhai P, Pörtner H-O, Roberts D, Skea J, Shukla PR, Pirani A, Moufouma-Okia W, Péan C, Pidcock R, Connors S, Matthews JBR, Chen Y, Zhou X, Gomis MI, Lonnoy E, Maycock T, Tignor M, Waterfield T (eds)]. Cambridge University Press, Cambridge/New York, pp 3–24. https://doi.org/10.1017/9781009157940.001

IPCC (2023) Climate change 2023: synthesis report. Contribution of working groups I, II and III to the sixth assessment report of the intergovernmental panel on climate change [Core writing team, Lee H, Romero J (eds)]. IPCC, Geneva, 184 pp. https://doi.org/10.59327/IPCC/AR6-9789291691647

Kuzma S, Saccoccia L, Chertock M (2023) 25 countries, housing one-quarter of the population, face extremely high water stress. World Resources Institute, 16 August 2023. Available online: https://www.wri.org/insights/highest-water-stressed-countries#:~:text=New%20data%20from%20WRI's%20Aqueduct,their%20entire%20available%20water%20supply. Accessed 2 Sept 2024

Lian X, Piao S, Chen A, Huntingford C, Fu B, Li LZX, Huang J, Sheffield J, Berg AM, Keenan TF, McVicar TR, Wada Y, Wang X, Wang T, Yang Y, Roderick ML (2021) Multifaceted characteristics of dryland aridity changes in a warming world. Nat Rev Earth Environ 2:232–250. https://doi.org/10.1038/s43017-021-00144-0

Mentaschi L, Duveiller G, Zulian G, Corbane C, Pesaresi M, Maes J, Stocchino A, Feyen L (2022) Global long-term mapping of surface temperature shows intensified intra-city urban heat Island extremes. Glob Environ Chang 72:102441. https://doi.org/10.1016/j.gloenvcha.2021.102441

Prăvălie R, Bandoc G, Patriche C, Sternberg T (2019) Recent changes in global drylands: evidences from two major aridity databases. Catena 178:209–231. https://doi.org/10.1016/j.catena.2019.03.016

Rillig MC, van der Heijden MGA, Berdugo M, Liu Y-R, Riedo J, Sanz-Lazaro C, Moreno-Jiménez E, Romero F, Tedersoo L, Delgado-Baquerizo M (2023) Increasing the number of stressors reduces soil ecosystem services worldwide. Nat Clim Chang 13:478–483. https://doi.org/10.1038/s41558-023-01627-2

Spinoni J, Barbosa P, Cherlet M, Forzieri G, McCormick N, Naumann G, Vogt JV, Dosio A (2021) How will the progressive global increase of arid areas affect population and land-use in the 21st century? Glob Planet Chang 205:103597. https://doi.org/10.1016/j.gloplacha.2021.103597

Tuholske C, Caylor K, Funk C, Verdin A, Sweeney S, Grace K, Peterson P, Evans T (2021) Global urban population exposure to extreme heat. PNAS 118(41):e2024792118. https://doi.org/10.1073/pnas.2024792118

Yuan J, Lin Q, Chen S, Zhao H, Xie X, Cai Z, Zhang J, Cheng T, Hua M, Zhang R (2022) Influence of global warming and urbanization on regional climate of megacity: a case study of Chengdu, China. Urban Clim 44:101227. https://doi.org/10.1016/j.uclim.2022.101227

Chapter 2
A Regional Planning Response to Water Scarce Cities: The Case of Monterrey, Mexico

Rob Roggema

Abstract In dealing with the increasingly arid conditions, and the scarcity of water in urban metropolises it is not sufficient to copy solutions that were useful in the past, when the problems were fundamentally different. The radical changes in climate conditions require a novel approach to keeping and creating water in those urban regions. In this chapter the regional planning opportunities for realizing this are presented. The first strategy is to think at the watershed level of the entire river basin and apply the three-step strategy. First, capture the water and let it infiltrate in the soil. Secondly, the water needs to be retained in basins to keep it as long as possible in the system. The final step is to create space for the water in urban contexts where the discharge of the water is slowed down. A second strategy beholds the case of water being absent. In this case, a more radical planning approach is proposed to create rain (water). Using the biotic pump principle of letting forests emerge and expand, to create their own rain, the rainfall and humidity in a region can substantially increase. Both strategies shall be applied in coherence, so the water supply of large urban conurbations can be kept up to level.

Keywords Droughts · Arid cities · Monterrey · Mexico · Regional planning · Landscape-driven planning

2.1 Introduction

Water scarcity in cities is no longer a minor issue. It is an increasing problem in many regions around the world. Though hypothetical, the movie ´Siccitá´ (Fasano 2022) shows us a potential scary future, in which a city (Rome) must deal with a prolonged drought, which lasts more than a year. People are desperate, the Tiber runs dry, the future looks hopeless and eventually people start rioting and wanting

R. Roggema (✉)
Escuela de Arquitectura, Artes y Diseño, Tecnológico de Monterrey, Monterrey, Nuevo León, Mexico
e-mail: rob@cittaideale.eu

R. Roggema (ed.), *Dry Urbanism*, Contemporary Urban Design Thinking, https://doi.org/10.1007/978-3-031-77347-1_2

to kill each other. Although this is a utopian, or rather dystopian, image of the future, large parts are not too far of reality. The movie offers a mirror to our current cities and what might happen if the trend of rising temperatures continues. The movie asks therefore implicitly to act. Take measures and invent solutions that will prevent the shown future form becoming a reality.

To deal with the lack of or scarce water resources, many different measures and solutions have been proposed. From saving water in households, construction of desalination plants and water recycling to measures that reduce the heat, such as planting trees or creating green spaces, or the construction of novel water infrastructures to transport water from far away to drought prone settlements.

At the regional scale urban agglomerations can also be planned in a way so the region can contribute to water conservation and/or -supply. In this chapter the Mexican context for the water problem will be sketched and possible planning strategies will be elaborated in the context of the Monterrey region.

2.2 Where Are We at?

Fresh water is a scarce resource. Of all water on earth 97.5% is salt water. Of the remaining 2.5%, only 0.3% is liquid fresh water (USGS 2013; Shiklomanov 1993). Moreover, this scarce resource becomes even more precious if the lack of good quality drinking water becomes apparent. Many countries lack sufficient drinking water for a large portion of their population. Mexico, for instance, scores 50 on a scale to 100 to drink water from the tap safely (Wall 2023, Yale Centre for Environmental Law & Policy 2023). Both the water quality and scarcity have been mapped at the global level (WWF Risk Filter, n.d.-a, b) and shows there are large parts of the globe that face problems with both or either of those topics. Mexico suffers both from a lack of water as well as a low quality.

Climatic change has driven this problematic situation, in combination with degradation of forests and loss of tree cover (Global Forest Watch n.d.). The heat, and because of this, increased and prolonged droughts, impact urban areas, nature, water availability and agriculture. In a large part of the country, summers are exceeding 40 degrees (CICESE 2023), and precipitation has declined significantly in the American Mexican region (Wilder et al. 2010). The droughts are widespread across Mexico (NASA Visible Earth 2021), and extend beyond a simple reduction in rainfall, but are increasingly leading to drying out of groundwater aquifers (Patel and Tierney 2022).

For the Monterrey-region, in the northeastern part of Mexico, the water problems have become visible in the decrease of the size and capacity of the drinking water basins (Fig. 2.1), but also in the social unrest that resulted from a sharp decline in water availability for the city's residents. For example, people have sabotaged water pipes (García 2022), and drivers of the trucks delivering water have been kidnapped (Tsui and Ibarra Chaoul 2022).

Fig. 2.1 Shrinkage of the capacity in La Boca, the Presa Rodrigo Gómez Reservoir in 2021–2022. (Source: Patel and Tierney 2022)

The situation in Mexico is very serious. Between 12 and 15 million people do not have access to drinking water, ranking as one of the top 25 countries in the world having to cope with water stress (Kuzma et al. 2023). 80% of the country faces desertification (Romero-Sanchez et al. 2016). The government of Mexico has launched an ambitious program of reforestation, Sembrando Vida, but first analyses show this program has not been able to counteract loss of forest cover. On the contrary, in 2019 a net amount of more than 72,000 hectares has been lost (Warman et al. 2021).

In March 2023, looking at all dams in Mexico the total capacity was 59% of the total, due to low rainfall (Conagua 2023), which poses a further risk to the water supply of large parts of the population. Rainfall is insufficient in more than half of the country (Ortega-Gaucin et al. 2018). The civil engineering of the urban water system is often both hidden underground (Gutiérrez 2023) as well as in bad condition, so a lot of water is leaking or evaporating without replenishing the aquifers. For instance, 40% of the water in Mexico City is leaking away (Hadley 2021). Improvements to the water treatment system cannot build on sufficient financial support (Mares 2023).

The lack of water, the water stress, the lack of investments in water systems, its invisibility and counteractive spatial planning actions, are, on the one hand side, exaggerated by climate change, on the other hand they are contributing to changes in climate. The worrying part here is that at the global level, the realization of the Paris agreement goals is far off track (United Nations Environment Programme

2023). This means that at the regional and local scale the problems will not decrease but rather intensify because of ongoing climate change. Appeal for local and regional actions to conserve water, care for water and provide the conditions for supplying the water demand in urban metropolises and centers.

It becomes increasingly clear that we have a large and urgent problem. The droughts are getting more serious, and it is very doubtful whether there will be enough water to provide to all human and non-human systems. Therefore, we need to develop fundamentally different solutions. Moving forward on the way we have done so far will not suffice. The planning has been dominated by a strong belief in technical, measurable outcomes. These have, in general exaggerated the problems. Especially when circumstances alter, the solutions from the past become redundant. Besides this, incremental changes will not be powerful and quick enough to cope with or solve the major problemswe face as humanity. To raise awareness about the preciousness of the water we have in the city, the first opportunity is to make the waterways and streams visible again. Most of the water is invisible, or neighborhoods are built with their backs to the streams and creeks. This way urban dwellers cannot relate to the water, let alone they feel the urgency to take care of the water in their environment. Moreover, the scope of solving problems in the current urban water systems is solving the urgent problem of today; stop the system from leaking, discharging excess water out of the way and preventing water from entering the urban neighborhoods. This focus solely on the short term, hinders the structural transformation towards a regenerative regional water system that is resilient, and can provide the amounts of water that are needed in an urban, drought prone, region.

2.3 What Can We Do?

The dry conditions of many urbanized regions around the world demand novel approaches. The applications of solutions that have been successful to solve past problems, will cause a mismatch when applied to the radical problems and conditions of the future. This will be causing the problems to exaggerate rather than being relieved.

The development of truly regenerative systems (Reed 2007; Wahl 2016; Roggema 2022), the mindset requires to move from the increased use of energy to allow urban system functioning to an operating system that gains energy. This way the degradation of the environment can be regressed and turned into regeneration (Fig. 2.2). For the water system it implies a shift from water used for human consumption in a technical, efficient system, towards a reciprocal system, returning more water of higher quality after use back into the environment.

As Fig. 2.2 shows, a regenerative system goes far beyond being ´green´ or ´sustainable´. Regeneration literally returns more (clean water) than it uses (in our urban systems) out of the environment (Roggema 2019). In the context of increasing droughts in urban agglomerations the regional planning and design must therefore altered to keep water in the area. Water must be treated as a precious resource

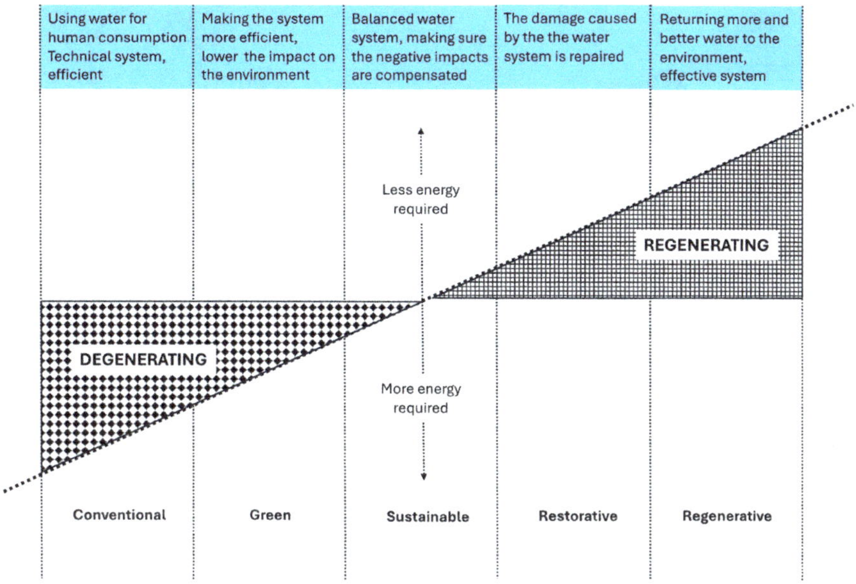

Fig. 2.2 From degenerating to regenerating systems. (Kumar et al. 2024; adapted from Reed 2007)

instead of a utility. For planning purposes, the most effective approach is to make use of a four-step-strategy: re(d)use-capture-retain-discharge or, in Dutch *besparen-opvangen-bergen-afvoeren*, which is an adaptation of the three-step-strategy as proposed by the Committee on water management in the twenty-first century, which consisted of capture-retain-discharge only (Commissie WB21 2000).

So, before planning a regional water system that conserves and retains as much water as possible, the first gain can be made by saving on water use. This can be applied within the household and in public spaces, where household (grey and black) water can be purified. This way the used water is brought back in the urban water cycle at the lowest possible scale. When this is executed at the lowest possible scale (house or in the nearest public space), it gives residents the awareness and responsibility to take care for the water and limits further extraction of water from natural resources. The neighborhood of the Paddock in Castlemaine in Australia illustrates the potentiality of this solution (Hes and Overall 2018), providing a clean waterbody close to the houses in a dry landscape (Fig. 2.3). It adds to the quality of the urban environment with both benefits to the residents (cooling, recreation, view), as well as for non-human organisms (ecological habitat, available water of better quality, space to live).

For regional planning, the three remaining steps are more relevant. The goal of keeping the water as long as possible in the system requires planning interventions at the regional watershed scale, from the moment the water enters the system, as rain or groundwater. Instead of discharging water as quickly as possible, and is lost,

Fig. 2.3 Purification of household water in the public space, The Paddock, Castlemaine. (Source: https://sustainablehouseday.com/listing/the-paddock-eco-village/)

from and through the city, the objective is to capture and store water before a slow discharge is considered.

As an example of this approach the proposed plan for the Santa Catarina River in Monterrey, Mexico (Roggema 2023) is presented (Fig. 2.4).

The upper part of the river basin, which is higher up in the mountains, must undergo several adjustments to capture the water and to allow it to infiltrate in the soil to supply local and regional natural aquifers where it is possible given the soil constraints. To allow the water the time to enter the soil horizontal ponds, which are surrounded by additional forests are designed in the upper parts of the watershed.

The water flowing out of this section is retained when the river flows to the flatter parts of the system. Large retention basins are proposed to store the water, which can be used for drinking water purposes, ecological reserves, or recreation. Here, larger amounts of water are stored for a long time.

In periods of heavy rain, and in the case of Monterrey, in the rare case of a hurricane, excess water will flow into the urban part of the river system. In the main river, not only excess water from the upper basins needs to find a place, but also the water flowing off the paved urban spaces adds up to the space required to slow down the discharge. The current concrete riverbanks operate as a canal, increasing the pace of discharge hence causing floods and damage in the urban areas. In this urbanized part of the river, principles derived from the room for the river program (Van

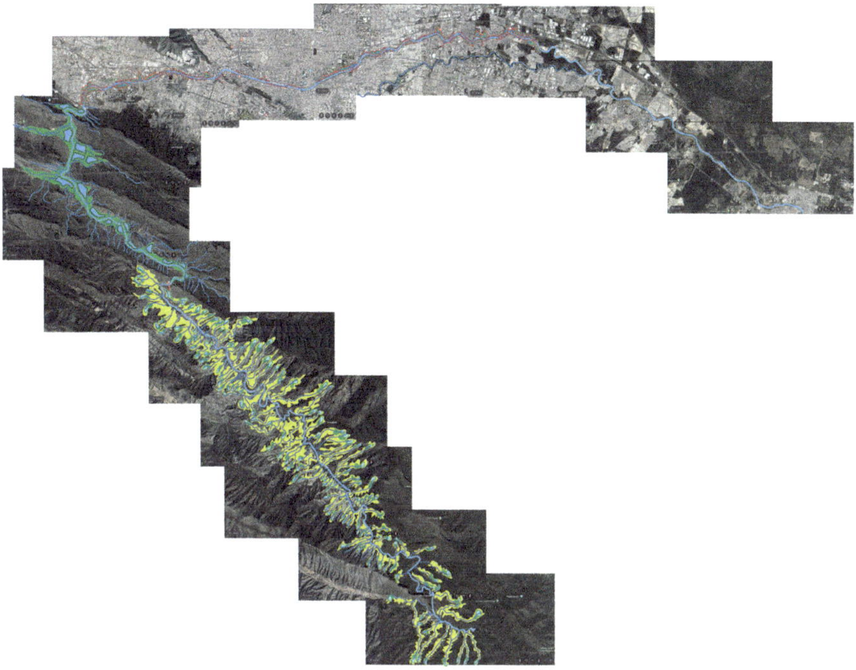

Fig. 2.4 Santa Catarina basin plan. (Source: Roggema 2023)

Alphen 2020) are applied. In spaces on the banks of the river, new wetlands, green river parks and the riverbed widenings are designed, to give the water a place to go, stay there for a while and be released after the main rainfall ended. The time the water stays in this part of the river is extended without flooding streets, squares, or houses. Current empty, un-, or underused spaces can become part of aspatial system that is more flexible. When there must be dealt with a lot of water these spaces can function as a temporary storage which are released when the system can handle it. Large parking lots, playgrounds, green wastelands, or subterranean cellars can be designed so the excess water from the river can enter it without flowing back into the riverbed at once. Small adjustments, such as an opening pathway, would be sufficient to make this work. The spaces themselves are available any time.

The capturing retaining and slowly discharge of water can therefore be planned for at the regional watershed scale. For each of its parts specific design strategies must be implemented, but the common denominator is that each of the design measures adds to the ecological quality, the resilience of the system, and the availability and quality of the water.

2.4 What to Do When It Stops Raining?

The three-step-strategy works well when there is (excess of) water. But recent conditions show there are also long periods without any or only little precipitation. In these arid conditions, water is absent, and this causes additional problems. From an empty plate one cannot eat. In this context a completely different alternative should be considered.

The question here is how (rain)water can be created. Apart from doing a Raindance, there is an experimental alternative option, making use of the biotic pump principle (Makarieva and Gorshkov 2007; Makarieva et al. 2013). The biotic pump could create higher levels of humidity and stimulate forest forming and rainfall. The principle emphasizes the stimulation of natural water flows in forested areas. It starts with planting the first hectares of forest near a large water source, such as an ocean or sea. In the case of Monterrey this is most likely the Gulf of Mexico. The planted forest attracts the evaporated water from the sea and uses it to grow. It will not only add to the growth of the existing trees but foresees also the expansion of the forested areas. New trees will be growing at the edges of the firstly planted part. The forest has then grown in size, which allows it to attract even more water from the sea. The forest evaporates the attracted water, causing rain, which is replenishing the forests itself. This causes new expansion of the forest, and a slow movement of forest inland will start. This process will continue, eventually growing a regional forest, similar to rainforests in tropical or subtropical regions.

The proposed plan for the Monterrey region (Roggema 2023), is to start at the Gulf of Mexico, where the Rio Bravo ends in the Gulf. The plan is to make use of the Laguna Madre y Delta de Rio Bravo, for the supply of water to the first forest. Simultanuously, first forests are planned upstream from the Presa's Marte R. Gómez and la Cuchillo. From each first forest the growth will move upstream (Fig. 2.5),

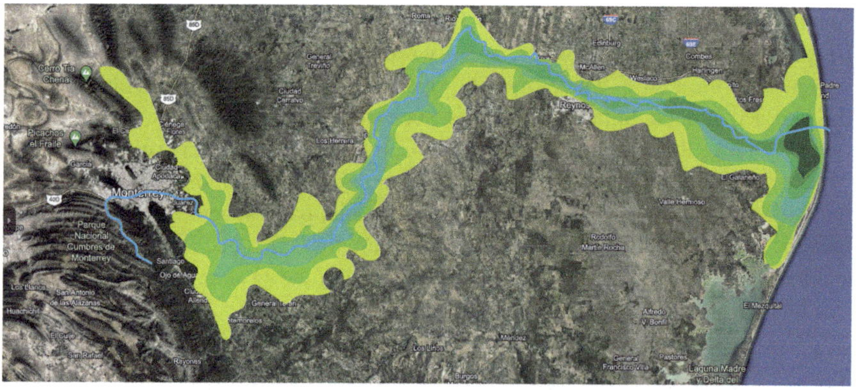

Fig. 2.5 Application of the biotic-pump principle to the Tamaulipas region, Mexico. The emergence of a natural regional forest that could provide humidity and rain to desert-like environments. (Source: Roggema 2023)

until it will reach the outskirts of Monterrey Metropolitan region, where the forestation will spread in the urban fringes around the agglomeration.

Growing forest and increasing rain is an experimental way to create higher levels of humidity and reciprocal rainfall. This potentially could add a serious amount of rain hence water to arid urban region. At the same time the forest could decrease temperatures and the Urban Heat Island effect and enhances biodiversity. This solution to create more water, must be applied at the large scale, and will only perform results on the longer term, as the forest will influence the regional climate. For Monterrey, it is likely necessary to think systemically, and connect the metropolis to the Gulf, enhancing the local climate conditions along the way.

2.5 Lessons Learned

The problem of arid regions is ongoing and will influence many urban areas across the globe. The current solutions to harness water, or recycle water are needed, but will not provide a structurally novel solution that is required to deal with the fundamental new climate conditions we face around the world. A regional planning approach could help to think differently, on the long term and the large scale. Only this perspective can be instrumental in achieving novel pathways to balance the climatic and ecological systems in dry areas.

For a spatial solution to be effective, two principles are promising: the application of the three-step-strategy, (capture-retain-discharge) at the watershed level and the biotic pump, establishing the emergence of natural forests, adjusting the climatic conditions. These two principles should be embedded in regional planning strategies. This starts with a vision of how the systemic change should be seen on the long term for a large region. Secondly, this large-scale plan must be implemented at smaller scales, in a simultaneous manner. Several states and municipalities need to be responsible to implement parts of the overall view, and to collaborate with each other. Water provision and availability is not a local matter but requires organizations at all scales to jointly develop the vision, designing the plan and implementing spatial projects.

References

CICESE (2023) Inicio de un caluroso verano 2023: dos ondas de calor y ¿contando? https://ulp.cicese.mx/inicio-de-un-caluroso-verano-2023-dos-ondas-de-calor-y-contando/. Accessed 17 Jan 2024

Commissie WB21 (2000) Waterbeleid voor de 21e eeuw. Geef water de ruimte en de aandacht die het verdient. Advies van de Commissie Waterbeheer 21e eeuw. Ministeri van verkeer en Waterstaat en unie van Waterschappen

Conagua (2023) Informe semanal del Comité Técnico de Operación de Obras Hidráulicas. Comisión Nacional del Agua, 14 de marzo de 2023. https://www.gob.mx/conagua/prensa/informe-semanal-del-comite-tecnico-de-operacion-de-obras-hidraulicas-328960?idiom=es. Accessed 30 Jan 2024

Fasano ML (2022) Interview with paolo virzì: "dry"—(an almost) true story. The Italian Reve, 27 September 2022. https://www.theitalianreve.com/interview-with-paolo-virzi-dry-an-almost-true-story/. Accessed 17 Jan 2024

García J (2022) El norte de México se seca. El País, 18 July 2022. https://elpais.com/mexico/2022-07-18/el-norte-de-mexico-se-seca.html#?rel=lom. Accessed 21 Jan 2024

Global Forest Watch (n.d.) Primary Forest Loss in Nuevo Leon. https://www.globalforestwatch.org/dashboards/country/MEX/19/?map=eyJjYW5Cb3VuZCI6dHJ1ZX0%3D. Accessed 21 Jan 2024

Gutiérrez IJ (2023) What options do we have to fight the water crisis? TEC science, 30 June 2023. https://tecscience.tec.mx/en/human-social/how-to-fight-the-water-crisis/. Accessed 30 Jan 2024

Hadley K (2021) Addressing water scarcity in Mexico City. The Borgen project. https://borgenproject.org/water-scarcity-in-mexico-city/. Accessed 30 Jan 2024

Hes D, Overall R (2018) The incredible challenge of water-positive development in rural Australia. The Fifth Estate, 8 February 2018. https://thefifthestate.com.au/columns/spinifex/the-incredible-challenge-of-water-positive-development-in-rural-australia/

Kumar M, Roggema R, Dogra S, Aguilar-Barajas I, Snyder SA, Furumai H, Mahlknecht J (2024) Global rise of thirsty cities: defining and redesigning the strategies for water sustainability. ACS EST Water 4:2766–3109. https://doi.org/10.1021/acsestwater.4c00362

Kuzma S, Saccoccia L, Chertok M (2023) 25 países, una cuarta parte de la población mundial, enfrentan un estrés hídrico extremadamente alto. World Resources Institute. https://es.wri.org/insights/25-paises-una-cuarta-parte-de-la-poblacion-mundial-enfrentan-un-estres-hidrico. Accessed 21 Jan 2024

Makarieva AM, Gorshkov VG (2007) Biotic pump of atmospheric moisture as driver of the hydrological cycle on land. Hydrol Earth Syst Sci 11:1013–1033

Makarieva AM, Gorshkov VG, Sheil D, Nobre AD, Li BL (2013) Where do winds come from? A new theory on how water vapor condensation influences atmospheric pressure and dynamics. Atmos Chem Phys 13:1039–1056

Mares F (2023) Mexico requires investment to treat more than 32% of wastewater. Mexico Businesss News. https://mexicobusiness.news/infrastructure/news/mexico-requires-investment-treat-more-32-wastewater. Accessed 30 Jan 2024

NASA Visible Earth (2021) Widespread drought in Mexico. https://visibleearth.nasa.gov/images/148270/widespread-drought-in-mexico/148270f. Accessed 17 Jan 2024

Ortega-Gaucin D, De la Cruz BJ, Castellano Bahena HV (2018) Drought vulnerability indices in Mexico. Water 10(11):1671. https://doi.org/10.3390/w10111671

Patel K, Tierney L (2022) Northern Mexico has a historic water shortage. These maps explain why. The Washington Post, 9 August 2022. https://www.washingtonpost.com/weather/2022/08/09/drought-mexico-water-monterrey/. Accessed 17 Jan 2024

Reed, B. (2007) Shifting from 'sustainability' to regeneration, Build Res Inf 674, 680, 35 (6); https://doi.org/10.1080/09613210701475753

Roggema R (2019) ReciproCity, giving instead of taking. Inaugural lecture. Hanze University of Applied Sciences, Groningen

Roggema R (ed) (2022) Design for regenerative cities and landscapes. Rebalancing human impact and natural environment, Contemporary urban design thinking, vol 5. Springer, Dordrecht/Heidelberg/London, 305pp

Roggema R (2023) Agua del Futuro: ¿Ser o no ser?. Keynote lecture at the National Water Convention, 22 November 2023, Monterrey, Mexico

Romero-Sanchez ME, Gonzalez-Hernandez A, Moreno-Sanchez F (2016) The assessment of land degradation and desertification in Mexico: mapping regional trend indicators with satellite data. In: Kaswamila AL (ed) Land degradation and desertification – a global crisis. InTechOpen [online]. https://doi.org/10.5772/64241. https://www.intechopen.com/chapters/51516. Accessed 21 Jan 2024

Shiklomanov I (1993) World fresh water resources. In: Gleick PH (ed) Water in crisis: a guide to the world's fresh water resources. Oxford, Oxford University Press

Tsui K, Ibarra Chaoul A (2022) Drought in Mexico leads to water rationing, theft. The Washington Post, 3 August 2022. https://www.washingtonpost.com/world/2022/08/03/mexico-drought-water-shortages-theft-rationing/. Accessed 21 Jan 2024

United Nations Environment Programme (2023) Emissions gap report 2023: broken record—temperatures hit new highs, yet world fails to cut emissions (again). Nairobi. https://doi.org/10.59117/20.500.11822/43922

USGS (2013) The world's water. https://web.archive.org/web/20131214091601/http://ga.water.usgs.gov/edu/earthwherewater.html. Accessed 17 Jan 2024

Van Alphen S (2020) Room for the river: innovation, or tradition? The case of the Noordwaard. In: Hein C (ed) Adaptive strategies for water heritage. Springer, Cham. https://doi.org/10.1007/978-3-030-00268-8_16

Wahl DC (2016) Designing regenerative cultures. Triarchy Press, Axminster

Wall K (2023) Where is the world's most dangerous drinking water? QS Supplies. https://www.qssupplies.co.uk/worlds-most-dangerous-drinking-water.html. Accessed 17 Jan 2024

Warman J, Zuñiga JI, Cervera M (2021) Análisis de los impactos en las cobertyras forestales y potencial de mitigación de las parceles del programa Sembrano Vida implementadas en 2019. WRI México. http://movilidadamableorg/WRIMexico/WRI%20México%20Análisis%20sobre%20los%20impactos%20ambientales%20de%20Sembrando%20Vida%20en%202019pdf. Accessed: 30 Jan 2024

Wilder M, Scott CA, Pablos NP et al (2010) Adapting across boundaries: climate change, social learning, and resilience in the U.S.–Mexico border region. Ann Am Assoc Geogr 100(4):917–928. https://doi.org/10.1080/00045608.2010.500235

WWF Risk Filter (n.d.-a) Water quality. https://cdn.kettufy.io/prod-fra-1.kettufy.io/images/riskfilter.org/WRF_2021_GLO_13WaterQuality_Light.png. Accessed 21 Jan 2024

WWF Risk Filter (n.d.-b) Water scarcity. https://cdn.kettufy.io/prod-fra-1.kettufy.io/images/riskfilter.org/WRF_2021_GLO_11WaterScarcity_Light.png. Accessed 21 Jan 2024

Yale Centre for Environmental Law & Policy (2023) Unsafe drinking water. https://epi.yale.edu/epi-results/2022/component/uwd. Accessed 17 Jan 2024

Chapter 3
Importing the Desert into the Urban: Architectural Narratives from the Middle East

Ali A. Alraouf

Abstract Multiple perceptions about the desert suggest that it is an empty space. Hence, no real valuable contribution can be extracted from calling for a deeper understanding of the desert. This chapter argues for an alternative perception where the desert is seen as not only an appropriate context for urbanism but also a source of inspiration for architects and city designer working in the Middle East. This chapter aims at initiating a discussion concerning desert development including developers and environmentalists to planners, community leaders, and those who dwell in desert regions. This chapter contests the significance of the desert in the contemporary architectural and urban environment in the context of Middle East and Gulf. The contestation is shedding light on the different approaches to the exploitation of the desert in creating a spectacle within the fabric of the unprecedently urbanized city. The chapter investigates different desert ecologies and the various ways in which they are used, misused, and navigated by the different actors. Using the trilogy of the desert, museum and contemporary Gulf cities, this chapter argues that with the current pace of development and urbanization, a new perception of the desert as a developmental context is needed. It concludes with a conceptual approach to articulate the role of the desert as a host for architecture and urbanism and as an inspirational matrix for urbanizing contemporary Middle Eastern cities.

Keywords Desert architecture and urbanism · Middle eastern cities · Contemporary Gulf cities · Museums studies · Doha, Qatar · Nomadism

3.1 Foundational Introduction

This chapter contests the significance of the desert in the contemporary architectural and urban environment in the context of Middle East and Gulf. The contestation is shedding light on the different approaches to the exploitation of the desert in

A. A. Alraouf (✉)
Hamad Bin Khalifa University, Doha, Qatar

R. Roggema (ed.), *Dry Urbanism*, Contemporary Urban Design Thinking,
https://doi.org/10.1007/978-3-031-77347-1_3

creating a spectacle within the fabric of the unprecedently urbanized city. In the domains of architecture and urbanism, reading the desert as a place is substantial. Desert, as a space of living, resilience and a manifestation of sustainability must be understood and analyzed. Intertwiningly, desert as a metaphor and imaginary source for contemporary architecture and urbanism should be included in a holistic analysis about desert as a space and a metaphor. The chapter investigates different desert ecologies and the various ways in which they are used, misused, and navigated by the different actors. Using the trilogy of the desert, museum and contemporary Gulf cities, this chapter argues that with the current pace of development and urbanization, a new perception of the desert as a developmental context is needed. An approach which, particularly in the fields of architecture and urbanism, transcends perceiving the desert as the manifestation of the notion of emptiness. Rather, desert should be seen as a context of development. Such position is conditional as any development in the desert must be based on a deep understanding of its natural laws and logic. So, desert, as many urbanists are arguing, can be a place to accommodate people in the fast-urbanizing world. The chapter contends that without knowing the value of the desert, its language, patterns of development, logic of life and resilience structure, any desert development will be doomed to failure.

> I have always loved the desert. One sits down on a desert sand dune, sees nothing, hears nothing. Yet, through the silence something throbs, and gleams... What makes the desert beautiful is that it hides an oasis somewhere.
>
> Antoine de Saint-Exupéry (2018), The Little Prince.

In the Middle East and Gulf States, deserts have long acted as spaces for traditional settlements designated for the protection of traveling nomads and a manifestation of indigeneity. Later, they were developed into desert towns connected with the portal settlement located around the Gulf waterfront to create an interesting form of urbanity where the desert and the Gulf are fluctuating between the attachment and detachment (Alraouf 2020). The aim of the chapter is to inspire discussion and contemplation for anyone interested in desert development, from developers and environmentalists to planners, community leaders, and those who live in desert regions. While different projects will be introduced and analyzed, the selected main case study is the National Museum of Qatar (NMOQ), designed by Ateliers Jean Nouvel and located on the waterfront of the capital city Doha. Qatar is an emerging nation in the Persian Gulf, a peninsula surrounded by water where the desert reaches into the Gulf. The Qatari descend from a nomadic Arabian people who settled in this maritime desert. Dedicated to the history of Qatar, the NMOQ is an architectural landmark of major historical and cultural value to the country. Contrary to the practice and discourse of looking at the desert from a development perspective, the chapter analyses the significance of the desert as a source of inspiration for visual art and architecture. The chapter aims, by using the case of National Museum of Qatar, to juxtaposition the engagement with the desert as a context and as a metaphor. Hence, several aspired values including identity, heritage interpretation, aesthetic references and economic geography can be better perceived and instigated.

> The desert is a natural extension of the inner silence of the body. If humanity's language, technology, and buildings are an extension of its constructive faculties, the desert alone is an extension of its capacity for absence, the ideal schema of humanity's disappearance. . . But the desert is more than merely a space from which all substance has been removed. Just as silence is not what remains when all noise has been suppressed. There is no need to close your eyes to hear it. For it is also the silence of time.
>
> Baudrillard, Jean (quoted from Lane 2000. p. 4).

3.2 On the Wisdom of the Desert

Desert is an essential component of the world geography. It is a space of knowledge production and dissemination. The context of the desert is an inspiring source of knowledge particularly in the fields of resilience, sustainability, and adaptability. Hence, a need to go beyond the spectacle of the desert is crucial to construct an alternative approach towards its understanding and utilization as a context of sustainable built environment. The desert as a natural context and the manmade settlements within it, construct outstanding examples of sustainability, community resilience, creativity and climate mitigation and adaptation. Examining the traditional desert settlements in the Middle East and Africa reveals a unique level of sensitivity and holistic understanding of the logic of life in the desert. As many researchers illustrated, the Bedouins have exhibited such deep understanding of the desert in constructing their settlements (Al Buainain 1999; Al Hathlul 1981; Nasser 2015). They have used local materials, compacted architecture, protective shells, and minimum resources (Fig. 3.1). Such process of development, as traditional cases illustrate in Egypt, Jordon, Algeria, Morocco, Yemen among other Middle Eastern States, were substantially missed in a lot of contemporary attempts to invade rather than blending and coping with the desert.

Moolla (2015) argues Al-Koni's "environmentalism" thus is an appreciation of the natural world which expresses itself through the desert as symbol, drawing on all the sacred traditions which have shaped the interlinked North African, Mediterranean, and Saharan regions. Al-Koni's fiction confirms in narrative form the insights of scholars of these traditions (Calleja 2014). Malloy et al. (2013), signals to the unavoidability of respecting the rich desert ecosystems. The pressing challenge is that development activities often resulted in transforming these desert lands of an ancient living landscape, to be replaced with one that is human-made and dependent on a large consumption of energy and natural resources.

> The desert is an open space, the desert is a naked soul, and the naked soul has laws other than the hidden spirit in the body or in nature. The miracle of the desert is that it is hardened, it is deprived from the plant, and may die many years, but it never dies its people.
>
> Ibrahim Al-Koni (2012). Legends of the Desert.

Throughout the Middle East, starting from the mid of the twentieth century, the modern state gradually emerged to end Bedouin life, marked the end of the myth of the desert. The State linked tribes that had previously only been associated with

Fig. 3.1 Siwa Oasis in Egypt as a manifestation of how traditional desert settlements blend with the environment and acknowledged the ecological conditions

their traditions and customs with new charters and regulations that they would not be able to resist and live outside for long (Nasser 2015). In short, the state came to dissolve what Ibn Khaldun called the micro-community, that is, tribal community, in a major one, the society of the state (Pribadi 2014). As a result, the multiple contemporary approaches to desert development neglected Bedouins and their unique way of life. Their traditions were exploited by tourism industry and transformed them into spectacle. In different desert around the Middle East, it can be arguing that Bedouins are the only human community wiped out by the modern state and its concept of development. The Bedouins could not adapt to the state. Before the State, they were the state themselves. They are the ones who make the law, define the sentence, and carry out the punishment (Nasser 2015). However, this is no longer possible with the State seeking to establish a social contract that the Bedouins were not familiar with before. Ceding part of their freedom to the state is new to them. But what seemed difficult, even impossible, happened with little resistance: the state was able to dissolve the Bedouins in its new container.

The significant feature about the concept of dwelling in the desert is centered around the notion of belonging. Such understanding is crucial in the discussion related to contemporary development projects in the desert primarily housing. This distinction between belonging and inhabiting is indicative. In a modern development discourse, the desert as infrastructure should be considered through these two generative notions. The traditional desert settlements in the middle east created a unique sense of belonging as they dealt with dwellers (Bedouins) as an integral owner of the development process as demonstrated in oases, towns, desert settlements. Paradoxically, the modern projects in the desert not only rejected or neglected the wisdom of desert settlements but also dealt with people as inhabitants. More painfully, the legal discourse of these communities about the desert use and repopulation vis-a-vis the state is unjust. All traditional forms of belonging and ownerships which are mainly related to unwritten agreements between tribal leaders, were rejected by the state. Such an unfair procedure resulted in the evacuation of different desert territories around the Middle East for the sake of new projects orchestrated by the state.

3.3 Contemporary Architectural Positions Towards the Desert

With three billion people estimated to be living in arid climate zones, this is a driver for designers and architects to develop more responsible infrastructure that works with existing natural resources. The desert dwellers used to be nomads, and the desert is so dynamic and changes constantly, while architects aim for stability and static interventions in territories. The discourse of architecture in the Middle East and particularly in relation to the context of desert was explored by many researchers (Bolleter 2019; Günel 2019). Yet, in most of the published work, a focus on the built environment and the physicality of places designed and built in the desert were emphasized. Hence, there are relevant questions to be investigated: How architects and urbanists look at the desert? As architects, what are architects and urbanists' contemporary position towards the desert? Would it be possible to go beyond the spectacle of the desert and analyze its unique context in a more profound manner? More significantly, throughout history, communities who dwelled in deserts are characterized with the highest levels of creativity, resilience, and sustainability. So, the question is how to look at the desert as a source of knowledge and platform for creativity? First and foremost, the desert is not only a vast piece of land as sometimes labelled in governmental statements and planning regulations around the Middle East and Africa. The desert is a sophisticated and compound context with social, cultural, environmental, and economical aspects rooted in eternal history as a main component of human geography and ecology. The rehabilitation of the desert should be done through sustainable measures which are manifested brilliantly in the traditional desert settlements dispersed and sometimes abandoned around the Middle East. A substantial consideration of local knowledge of the desert as a built and used environment by tribal and local communities that remained outside state structures for decades. There are some good examples of how buildings have been designed to "harness traditional wisdom with contemporary dependences".

The place of desert in the Arab and Gulf Architecture and Urbanism is a multifaceted narrative. In this section, I will crystalize architects' position towards the desert within the Middle Eastern context. Such analysis is based on evaluating the different conceptual approaches adopted by architects practicing in the Middle East and the Gulf. Therefore, a focus on the conception of desert in the modern built environment is crucial. For the context of this discussion, four main approaches can be identified:

1. The Fantasization of Nomadism
   In this approach, number of projects particularly for tourists' consumption, are articulated around the fantasies of Bedouins and Nomads' lives. The main features of such projects are compacted urbanism, tents, earth colors, and architectural vocabulary copied from traditional desert settlements. Projects like Al Maha resort in Dubai (Fig. 3.2) and Al Badayer in Sharjah are examples of this approach. Such gated resorts are a demonstration of emphasizing the sublime, mystical, re-consuming and remanufacturing the desert architecture.

   In the case of Al Maha Desert Resort in Dubai, the development is based on the concept of an oasis as the developers claim. The thorough examination of the project reveals that the value of the desert in the project is merely a context to emphasize the fantasization of the oasis and the Bedouins. Nothing of the Oasis' ecology or the wisdom of its built environment was authentically reflected in the project. Rather, the project is a sophisticated resort designated for tourists and affluent class of the society, in the form of a gated community which consists of a few scattered villas each of which enjoy the view to its adjacent freshwater swimming pool. Such projects distort the meaning of traditional desert settlements and the unique way of life of its dwellers (Fig. 3.3).

Fig. 3.2 The case of Al Maha Desert Resort in Dubai where the context of the desert was used as a means for tourism without respecting the logic of building and dwelling in the desert

Fig. 3.3 The case of Al Maha Desert Resort in Dubai is a mifistation of neglecting the essence of desert' sustainable wisdom as every unit is equipped with freshwater pool

2. Importing Modernity into the Desert

   This category is about the narratives of inserting the modern into the desert. Within this approach, architects will select to deconstruct the desert context by juxtaposition rather than blending. Hence, the architectural intervention is seen as a statement of complexity and contradiction within what might be considered for modernists as the boring and monotonous desert landscape. Importing modernity into the desert would give the needed spatial and visual excitement.

   Such a constructed mirage of the modern (the city) in the desert can be seen in the Maraya project in Al Ulla desert, Saudi Arabia (Fig. 3.4). This area became Saudi Arabia's first ever UNESCO World Heritage Site. The area features naturally carved rocks, historic writings, and cave art as well as many Islamic and religious monuments. The Royal Commission for AlUla (RCU) constructed this unique building in the deserts of AlUla and used mirrors in its exteriors to reflect the surrounding environment. The hall's design makes it an architectural extension of the environment surrounding it.

   Al Ula is a place of heritage for the world, and this is our 21st century realization of our vision to create a cultural hub, with Maraya as a stunning centerpiece venue for world-class events, performances, celebrations, and business gatherings. Maraya is a medium that intertwines nature, culture, and humanity. – RCU CEO Amr AlMadani, CEO of RU

Fig. 3.4 The *Maraya* concert hall in Al-Ula, Saudi-Arabia

The glistening square has earned them a Guinness World Records title for the largest mirrored building in the world, measuring 9740 Square Meters. The Maraya concert hall is named as such because 'Maraya' means reflection or mirror in Arabic. The 500-seat capacity hall was designed as a cuboid covered with mirrors externally. The building was designed to reflect and highlight the natural beauty of desert around Al Ula. The architectural art piece will serve as a venue for events, celebrations, performances, and business gatherings. The concert hall was built around *Mada'in Saleh* (which means Cities of Salih), a historic city which was known as a trade root connecting the Arabian Peninsula to the Levant region.

Another landmark in the influential projects importing the modern into the desert is the controversial Masdar city in Abou Dhabi (Fig. 3.5). The narrative of the city stimulates a fundamental question regarding the city's destiny: is it a sustainable bubble or a ruptured dream? Contesting the credibility of Masdar City—Abu Dhabi as an urban utopia in the desert? In 2006 Abu Dhabi launched an ambitious project to construct the world's first zero-carbon city: Masdar City. The fascinating story of Masdar demonstrates the stark difference between vision and reality (Fig. 3.6). Masdar City was fascinating, because its planners not only offered insights about urban design, but also proposed innovative ways of imagining energy and climate futures. Designed by London-based architecture office Foster + Partners, the eco-city cost twenty-two billion dollars and would house fifty thousand residents and forty thousand commuters on a six-hundred-hectare site. Günel (2019) described Masdar City as "a mirage that falls apart as you get closer to it," and argued that the technologies Masdar promoted did not actually work. Finally, he concluded, Masdar is "a lesson for how delusions of grandeur, technical mistakes, and above all poor planning can rob a fascinating idea of its credibility."

Fig. 3.5 The vison of Masdar as created by prominent architect Norman Foster as contemporary sustainable utopia (Foster and Partners)

Fig. 3.6 Sustainable urban planning in Masdar City, Abu Dhabi. (© Jan Seifert, CC BY 2.0, https://commons.wikimedia.org; https://commons.wikimedia.org/wiki/File:Masdar_city_under_construction_2012_crop.jpg)

In Spaceship in the Desert, Günel (2019) examines the development and construction of Masdar City's renewable energy and clean technology infrastructures, providing an illuminating portrait of an international group of engineers, designers, and students who attempted to build a post-oil future in Abu Dhabi. While many of Masdar's initiatives—such as developing a new energy currency and a driverless rapid transit network—have stalled or not met expectations, Günel (2019) analyzes how these initiatives contributed to rendering the future as a thinly disguised version of the fossil-fueled present. Spaceship in the Desert tells the story of Masdar, at once a "utopia" sponsored by the Emirati government, and a well-resourced company involving different actors who participated in the project, each with their own agendas and desires.

Utopia versus reality: Despite the theoretically promising first signs, experts are skeptical whether Masdar City can function beyond the drawing board. Energy efficient structures need to be created first which can also be used for people with low incomes so that the eco-city is also filled with life at the weekend too. At present, Masdar City is geared towards commuters who do business here and then get into a plane to head back home. This contradicts the aim of an emission-free city and makes a nonsense of the concept. Nonetheless, Masdar City serves as an inspirational example, which, on a rudimental level at least, shows how high technology and research ensure more sustainability. Whether the ambitions of the project will ever be realized remains to be seen.

3. Desert as a Land rather than a Context.
   In this approach and as evident in several countries around the Middle East, the desert is seen as a vacant land not a unique natural, geographical, and social context. Hence, this approach deals with the desert as a venue for solving development issues and challenges primarily in the housing sector. Projects targeting the poor and marginalized or projects, mainly gated communities, targeting the elite and rich, would be erected in a total absence of lessons learned from the traditional architectural and urbanism in Middle Eastern deserts. The required change here stems from the fact that desert is not a simple vacant land or an empty city block. Built evidence around the Middle East showed that development projects in the desert without a deep understanding of its logic, resulted in failing projects.

   The case of Egypt's recent developments is a manifestation of neglecting the contextual understanding of the desert. Egypt has placed its hopes on developing its vast and empty deserts as the ultimate solution to the country's problems. New cities, new farms, new industrial zones, new tourism resorts, and new development corridors, all have been promoted for over half a century to create a modern Egypt and to pull tens of millions of people away from the increasingly crowded Nile Valley into the desert hinterland (Sims 2015). The results, despite colossal expenditures and ever-grander government pronouncements, have been meager at best (Fig. 3.7), and today Egypt's desert is littered with stalled schemes, abandoned projects, and forlorn dreams (Sims 2015). It also remains stubbornly uninhabited due to the lack of contextualizes urban expansion in the desert within the bigger desert development understanding.

Fig. 3.7 As Cairo spreads outwards onto the agrarian fields, areas such as Ard el-Lewa are nearly fully urbanized. When the market reaches saturation point in the city, new developments seek empty land to build speculative housing on. Large construction sites continue to break desert ground in Greater Cairo. (Credit: Courtesy of MAS Urban Design 2014–2015)

In some of the Middle Eastern Countries, development of the desert particularly for housing projects is characterized by building for low income or displaced community suffering from urban gentrification. As Allen (2019) argues, the housing crisis in the Middle East is frequently confronted by pushing lower income groups towards the desert in public housing projects which imposed on the desert and have learned nothing from the traditional settlements. Allen (2019) argues that these projects lack plenty of community services primarily the absence of public transit creates barriers to viable employment and accessibility to cultural networks and plays a role in increasing social inequality. Allen (2019) investigates how housing and transport policy have played their role in creating these "Transit Deserts," and what impact race has upon those likely to be affected.

4. The Artificiality of the Desert in the City (the Modern).
This approach focusses on attempts to use the city as a platform to export the desert. In some projects, within Gulf cities and beyond, we can notice the process of bringing the desert into the city by creating a shallow dialogue between modernity and authenticity within the same building or project. Such attempts resulted in visual collage where the desert and its unique architecture were introduced to the city urbanity only visually and not via understanding the real values of desert's architecture and urbanism. In this approach architects and urbanists try to create a dialogue within the contemporary city between its modernity and vocabulary generated from the desert environment. Different architectural statements within the city managed to bring the desert into the modern via creating an architectural dialogue within the same projects. A typical case would be using two architectural languages within the same building, one represents authenticity of the desert, and one represents the modern rhythm. The justification for such a visual dialogue is always stressing the notion of assemblage identity. The rational is the city is modern and contemporary but has its own roots which might be visualized in the images of its projects.

The notion of bringing the desert into the city should move beyond the shapes and visual elements. It is interesting to consider the context of the desert as form inspiring platform. For example, sand dunes or patterns of sand movements can inspire architects and designers. Yet, bringing the desert into the city should be more holistic and with the aim of creating equally sustainable living environment. More significantly, being inspired by the desert and bring it into the city's urbanity invite us to focus on urban sustainability and how to design urban systems that provide a secure future for community development particularly in water security, sustainable building practices, and bold architecture and community designs (Malloy et al., 2013).

3.4 Contemporary Gulf Cities Transformed

Hotter and dryer than most parts of the world, the Middle East could soon see climate change exacerbate food and water shortages, aggravate social inequalities, and drive displacement and political destabilization (Rabinowitz 2020). And as renewable energy eclipses fossil fuels, oil rich countries in the Middle East will see their wealth diminish. How oil-rich economies prepare for a future with less oil needed as the world is moving swiftly to a more sustainable future characterized by less carbon emissions, mitigating climate change and relaying on renewable sources for needed energy? Amidst these imminent risks is a call to action for regional leaders. Could countries such as Saudi Arabia, Kuwait, and the United Arab Emirates harness the region's immense potential for solar energy and emerge as vanguards of global climate action? The Power of Deserts surveys regional climate models and identifies the potential impact on socioeconomic disparities, population movement, and political instability. However, offering more than warning and fear, (Rabinowitz 2020) highlights a potentially brighter future a recent shift across the Middle East toward renewable energy.

The last decade witnessed an unprecedented pace of development which almost characterized the urban scene in contemporary Gulf cities. Economic growth in the Gulf States of Kuwait, Bahrain, Qatar, UAE, Oman, and Saudi Arabia is happening at extreme speed, with results both familiar and spectacular: ultra-modern cities, major new tourism destinations, and unprecedented global economic engagement through huge sovereign wealth funds investing worldwide. Mirroring this development, the last five years have also witnessed extensive and rapid growth in museum activity, with the regeneration of existing museums, the founding of major new ones, and the creation of international satellites (Erskine-Loftus 2013).

3.4.1 The Authenticity of Sustainability: Contesting Dubai

Bolleter (2019) evaluates the designed landscapes of Dubai against the principles of sustainability, climate change impact and desert adaptability, and argues that what is happening in Dubai represents a significant discrepancy between theory and practice. She explores how designed landscapes can play a vital role in constructing a city's global image and legitimizing its socio-political hierarchy. Particularly those with an interest in Dubai or the many cities in the region that are experiencing Dubaification.

> Water, water, water.... There is no shortage of water in the desert but exactly the right amount, a perfect ratio of water to rock, water to sand, insuring that wide free open, generous spacing among plants and animals, homes and towns and cities, which makes the arid West so different from any other part of the nation. There is no lack of water here unless you try to establish a city where no city should be.
>
> Abbey, Edward. (1990).

Bolleter (2019), in surveying the landscape of Dubai, argues that the Emirate's rulers employ a "paradisiacal image" of greening the city. Dubai was viewed by Western researchers as being a "fantasy world in the desert" (Davis 2007, p. 63). Yet, its impact as an urban development model is manifested regionally and globally. The critical examination of Dubai's model of creating an influential urban development blueprint reveals significant issues. The pace of development in Dubai neglects the logic of desert development. The model has created an unsustainable city that consumes vast resources of desalinated Gulf water and energy. Also, the desert and Gulf ecosystems were significantly damaged in the process of creating the desert fantasy with its glitzy towers (Kolo 2016; Bolleter 2019; Alraouf 2020). Rizzo (2014) argues that Dubai's model was disseminated to Gulf cities, Middle Eastern Cities and cities in China, India, Brazil, Baku, etc. The spread of Dubai's model to the region and beyond was the outcome of regional interventions of gargantuan development companies, such as Emaar and Dubai Holdings. Alternatively, presidents, leaders and rulers in other countries were so much influenced by the image created and wanted to duplicate it in their own territories.

3.4.2 Contemporary Museums in the Gulf

In the last decade, the Gulf States have activated the most explosive museum building boom per capita in history. Why Building Museums in the Gulf? Museums in the twenty-first century are no longer just buildings belong to the cultural categories. Museums became cultural, social, educational and entertainment institutions blending with the city and the community. It has become a major tool in the marketing of knowledge and creative cities. Over the past six decades, Arab nations have taken increasing control not only over their own destiny and sovereignty but also on their cultural aspects. Part of which is the dramatic change in the fields of archeology, cultural and architectural heritage, museums, and cultural tourism. Wakefield (2020) offers new insights into the history and development of museums within the region. Drawing on regional and international scholarship, the book provides a critical and detailed analysis of museums and heritage institutions in Bahrain, Jordan, Kuwait, Oman, Qatar, Saudi Arabia, the UAE, and Yemen. Questioning and engaging with issues related to the institutionalization of cultural heritage, contributors provide original analyses of current practice and challenges within the region to

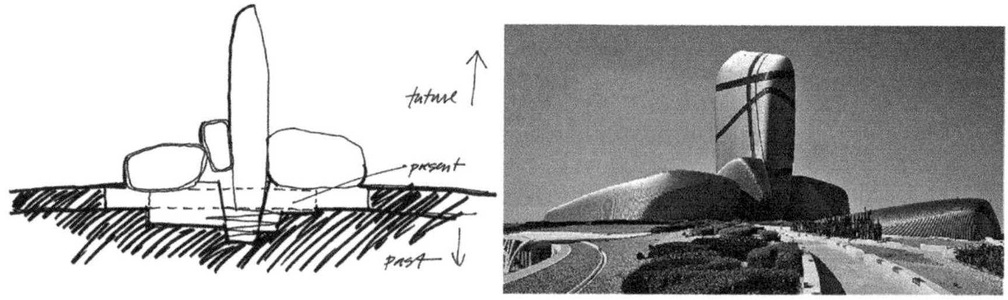

Fig. 3.8 King Abdulaziz Center for World Culture—Ithra—Dharan was designed by award winning Norwegian architectural firm Snøetta marked a shift in Gulf Museums design from traditional to a contemporary formal statement inspired by the context (Snøetta, *Projects*, n.d. https://snohetta.com/projects)

examine how museums are actively produced and consumed from both the inside and the outside. This critical analysis also enables debates to emerge that question the appropriateness of existing models and methods and provide suggestions for future research and practice. Wakefield 2020 offers fresh perspectives that reveal how Gulf museums operate from local, regional, and transnational perspectives.

Primarily, in the Gulf cities, the notion of using the museums as a catalyst for branding the city is evident. A critical review of such a boom in constructing museums around the Gulf cities reveals a few controversial points. Specifically, since the beginning of this century, the use of museums as a form through which to create national identity has flourished in the Gulf states. As Erskine-Loftus (2013) argues, museums creation and building has grown extensively in the Arab peninsula, particularly in the UAE and Qatar. In such context, museums were perceived not only as a symbol of modernity but as a manifestation of transforming Gulf cities into global destinations. A shift in using architecture as a tool to emphasize local identity took place in several Gulf cities. In the case of Saudi Arabia, the change of the architectural language used for building museums is significant and worth deep analysis. While the national museum was inspired by traditional Saudi architecture, King Abdulaziz Center for World Culture—Ithra—Dharan (Fig. 3.8) was designed by award winning Norwegian architectural firm Snohetta. "In the last decade, Abu Dhabi and Doha have embarked on a race to emerge as the region's cultural capital." (Alraouf 2015). Between Doha's multiple museum projects, including the I.M. Pei-designed Museum of Islamic Art, and Abu Dhabi's grand Saadiyat development, the cultural epicenter of the region may be shifting from traditional cities like Cairo and Baghdad to these emirates—although the political uncertainty in Cairo and Damascus could certainly have an impact on their success (Bharadwaj 2012).

> There is a misinformed tendency to think that the emphasis on the exterior will dilute the importance of what lies within. However, any art historian will validate that architecture is the most visible expression of a culture. Commissioning the greatest living architects of our time to house these collections is simply the way it should be". ("Cultural Jewels in the Gulf – The New York Times)
>
> M. Pocock, Art Gallery owner quoted from Bharadwaj (2012).

ddle
com-
small
Qatar
eera,
-level
d.

recog-
ue will
energy
culture
d. Qatar
map for
rld-class
e hugely
proaches
to create
in Doha
owledge-
dd to the
National

ıalization of Qatar and the Evolution of Doha

'a in the east of Arabian desert and covers an area of 11,437 km^2, 563 km. The country is a mostly low-lying and flat stretching into ands of Saudi Arabia. Until few decades ago, Qatar was domi- people whose livelihood depended on fishing and pearling (Al Kholaifi 2003; Alraouf 2022). However, the discovery of oil and ıot only socio-economic changes, but urban and environmental overall population witnessed an unprecedented increase since ation of the State's winning the bid to host the FIFA world cup pulation is 2.8 million, of which only around 15% are native Doha is home to more than 90% of the country's population, are professionals from all around the globe (Qatar Satistics) the mid-1960s, the majority of Doha's built environment ional houses that represent local responses to the surround- cultural conditions. During the 1970s and 1980s Doha was rnized city because of extensive use of oil revenues in mod- ularly housing, infrastructure, roads, and community facil- A glance at photographs of Doha in the 1950s and 1960s, . 3.9), is sufficient to understand how much this part of the a little village, it has become a capital.

ıning of the new millennium proved to be the turning era ted heavily in changing its image not only as a modern ic newcomer aspiring for a spot on the global stage 018a, b). In the last decade, Qatar managed to provide a

chit

Fig. 3.9 olution from a humble traditional settlement to a metropolitan urbanity. (Source: a stimulating, yet provocative, development model within the Gulf and the Mi East. As Roberts (2017) claims rarely has a state changed its character so pletely in such a short period. Previously content to play a role befitting its size, Qatar was a traditional, risk-averse Gulf monarchy until the early 1990s. diversified its foreign relations, established the satellite broadcaster Al Ja assumed a leading role in international mediation, and hosted several top sporting tournaments, culminating in the successful FIFA World Cup 2022 bi

3.6 Qatar Museums and Cultural Identity

Central to the Emir of Qatar, Sh. Hamad bin Khalifa Al Thani's vision is the nition that, as plentiful as Qatar's natural energy resources may be, their va 1 day be significantly diminished by the emergence of sustainable forms of production (Woodman 2008). The vision is to mold Qatar into a capital of and to make Doha a cultural bridge between the Middle East and the worl established Qatar Museums (QM) to be responsible for drawing the road many of the new cultural projects, primarily museums. The creation of a wo set of cultural facilities is a key element of the State's vision and includes th ambitious plan to build 10 completely new museums. QM's mission ap museums not as independent and isolated institutions—but rather working museums that are integrated into the urban spaces and lives of people living and the world (Alraouf 2016). As illustrated, within the new paradigm of kn based urban development in Doha, several museums were constructed to city's cultural landscape including the Museum of Islamic Art (MIA), Qata Museum (QNM) and the Orientalists Museum in Lusail.

3.6.1 An Analysis of the National Museum of Qatar in Doha (NMOQ)

> The National Museum is dedicated to the history of Qatar. Symbolically, its ar evokes the desert, its silent and eternal dimension, but also the spirit of modernit ing that have come along and shaken up what seemed unshakeable.
>
> NMOQ Architect Jean Nouvel

The National Museum of Qatar in Doha (NMOQ), for being the most plification of a verbal and direct metaphor. In this section, the National Qatar (NMOQ) will be analyzed and illustrated as an example of the m of the desert into the contemporary city fabric. The reflections on NM tecture and urbanism reveal that the Museum is inspired by a deep und the desert as a form, ecology, and concentrated social space. The NM ative contribution in the heart of the city balanced between old and and contemporary, authentic, and innovative and blending and iconic.

Fig. 3.10 The dynamic dialogue between old and new as NMOQ is built around the historic palace of the founder of modern QatarThe dialogue between old and new, heritage and contemporary and global and local as exhibited in the National Museum of Qatar. (Source: author)

1.5 million-square-foot site at the south end of Doha's Corniche, it will be the first monument travelers arriving from the airport will set their eyes upon. Conceived as growing out of the ground, the building uses rings of low-lying, interlocking pavilions, to encircle a large courtyard area and encompass 430,000 square feet of indoor space. Tilting, interpenetrating disks define the pavilions' floors, walls and roofs and the exterior in a sand-colored concrete (Fig. 3.10).

3.6.2 Contextualizing History, Embracing the Old Heritage

The first Qatar National Museum was established to display the elements of Qatar's history and the traditional modes of life of its people and inaugurated on the 23rd of June 1975 in the palace of Sheikh Abdullah Bin Jassim Al-Thani son of the founder of modern Qatar. The old Amiri Palace was the early twentieth century residence of the late Amir Abdullah bin Jassim Al-Thani. From 1972 to 1975, it was restructured and transformed into the first version of the Qatar National Museum. The project by Michael Rice and Anthony Irving won the Aga Khan Award for Architecture in 1980 (Rice 1977). The Old Palace comprises nine separate buildings. Sheikh Abdullah and his sons occupied the palace around 1901and used it as a residence and a ruling palace. It served as his family residence and the seat of government for approximately 25 years. In addition to the original Amiri Palace, the former Qatar National Museum included a Museum of the State, a lagoon, and a very popular marine aquarium. In 1980 the building won the Agha Khan award for restoration and rehabilitation of Islamic architecture. This splendid monument to Qatar's past is now preserved as the heart of the new National Museum of Qatar.

The museum narrative flows through the spaces just like the architecture of the building itself. Presented in 11 galleries, it's a 1.5-kilometer journey through three "chapters" of Qatari history: Beginnings, Life in Qatar, and Building the Nation. The organic design of the museum which consists of different sizes of desks allowed the new museum to literally embrace the old museum and intensify its value. The actual embracing of the old part not only created an interesting juxtaposition but also emphasized the appreciation of the old heritage. The masses of the new museum articulated around the old one gives a message of integration and appreciation rather than separation and discontinuity between old and new (Fig. 3.10).

3.6.3 The Desert Inspirational Role: Ecological Reinterpretation

The building's notorious shape is a gargantuan representation of the desert rose stone: a cluster of crystallized disks made of sand, baryte, and gypsum that occur in arid landscapes. According to Ouroussoff (2010), Nouvel's initial proposal was a hypogeal structure, but the project was not dubbed sufficiently representative of the country's ambitions, and the architect was invited to submit a more evocative shape. The rose, while not regarded as a local emblem before the construction, brought the desert into contention to support the museum's need for local character (Exell 2016, p. 66).

> bladelike petal of the desert rose, a mineral formation of crystallized sand found in the briny layer just beneath the desert's surface.
>
> Architect J. Nouvel.

The vast, 350-meter-long and the $434 million building was designed by Pritzker Prize-winning architect Jean Nouvel in the image of a desert rose (Fig. 3.11). The new museum's innovative design is inspired by the desert sand rose and grows organically around the former palace. Huge, sculptural disks, intersecting and cantilevering in all directions, surround the renovated Palace of Sheikh Abdullah bin Jassim. Nouvel, who was also behind the Louvre Abu Dhabi on Saadiyat Island, where Abu Dhabi is building its own cultural attractions.[1] The museum is a sprawling array of curved discs, cantilevered angles, and surprising intersections, and was

[1] Saadiyat will also be the home to the UAE's National Museum and a branch of the Guggenheim Museum designed by another renowned architect, Frank Gehry.

Fig. 3.11 The desert rose is also known as "sand rose" and "rose rock". Local nature provided an endless source of inspiration for J. Nouvel's QNM design

inspired by "desert rose" crystal clusters which form in the Qatari desert. It's an impressive marriage of past and present. Inspired by the desert rose with its interlocking disks, architect Jean Nouvel's innovative design, responds to the country's desert location by the sea (Jodidio, Philip 2020). The museum, with its connection to the original nineteenth-century palace, honors Qatar's heritage while looking to its future as a thriving cultural hub.

As stated by NMOQ architect, local nature provided an endless source of inspiration for his design. The desert context and the coastal culture are full of signs which need to be exploited for any creative journey. Precisely, the form of the desert rose, a unique feature of the Gulf coast, made a huge impact on the architect's quest for a local natural element to be reinterpreted geometrically to provide the formal composition of the museum. Hence, NMOQ's form is not a continuity of what a signature architect typically uses as evidence of his/her own ego. NMOQ is different in this sense as it is not representing the signature of Jean Nouvel in a new territory. Some quick revision of design trends used in similar project around the Gulf reveals a tendency to celebrate the architect's ego and signature at the expenses of respecting the locality.

The museum's design ability to bring the desert maturely into the city resulted in a provocative architectural language adding positively to the spectacle of the city. The museum's architecture and structure symbolize the mysteries of the desert's concretions and crystallizations, suggesting the interlocking pattern of the bladelike petals of the desert rose. The complexity of the tilted desert rose desks made of sophisticated steel structure and masterfully embracing the old palace with its traditional vocabulary, resulted in a provocative, stimulating and eye-opener series of external and internal visual, sensual, and spatial experiences.

3.6.4 NMOQ as an Engaging Museum

For developing new audiences for museums, a shift towards social inclusion is fundamentally needed. As Black (2005) argues, social exclusion is a main obstacle for creating engaging museums. Here, architecture can play an important role by providing spatial experiences which would include people from different cultural, social, and even ethnic backgrounds in the museum. The spatial experiences provided by QNM are based on the concept of interactivity and engagement of visitors. Technology is exploited to emphasize the interactive nature of the museum. Internal walls are used as screens to digitally narrate Qatar history and its evolution from a humble fishermen village to a global model of development and progress. Additionally, the museum outdoor and indoor spaces are articulated in a way that would create multi-sensory experiences for visitors. QNM utilizes technology to create a thoroughly new institution. Entire walls become cinematic displays and hand-held mobile devices guide visitors through the thematic displays of the collections.

Another aspect in QNM ability to engage results from its ability to create stimulating museum experiences by substantiating the role of visitors. QNM visitors are perceived not as negative receivers but positive engaging contributors to the museum collection and narrative. Perceiving the museum also as an educational tool where a culture of learning is intensified within its spaces helps embracing the concepts of social justice and cultural democracy (Hooper-Greenhill 2007; Black 2005). Engaging in the museum's lifelong learning experience would create a different mindset within Gulf communities and reestablish the role of museums in their lives.

3.7 Conclusions

For millennia the desert has been a harsh a difficult climate to build in. But in modern history, the nations of the Middle East have harnessed the power of the region to build great cities from the sand. Although most of the desert housing is prosaic, modern architecture was reinvigorated by the challenge of building in arid landscapes. Many of the most iconic examples of desert Modernism are found in the Middle East, owing once again to the people's creativity, knowledge, and resilience. Analyzing the traditional desert settlements and available technologies provides solutions to help promote ecological balance between nature and the built environment in the Middle East and beyond. ′Design Solutions That Work with the Local Environment′ can be concluded, assembled, and implemented in future projects and developments. In the case of Doha, Qatar, nomadic people build its capital city and talks about it through this emblematic monument built with the most contemporary construction tools (steel, glass, and fiber concrete), and will communicate through high-definition cinema, incorporating visitors' movements into its museography: this museum is a modern-day caravanserai. From there you leave for the desert, and you return from it bringing back treasures: images that remain forever engraved in your memory. The National Museum of Qatar is evidence of how intense this energy is. Of course, it will be home to the traditional geological and archaeological artifacts; of course, tents, saddles and the dishes will bear witness to nomadic life; of course, there will be fishermen's utensils, boats, and nets. Most importantly, though, it will spark an awareness that could only otherwise be encountered, experienced, after months spent in the desert, in pursuit of the particularities that elude our grasp except when the whims of time and nature allow. Everything in this museum works to make the visitor feel the desert and the sea. Qatar National Museum of Qatar (NMOQ) is part of such widely implemented interventions within the fabric of Doha. Such projects have been coordinated and fully integrated with the city to a balanced cultural identity in the era of knowledge economy and knowledge-based urban development.

References

Abbey E (1990) Desert solitaire: a season in the wilderness. Touchstone

Al Buainain F (1999) Urbanisation in Qatar: a study of the residential and commercial land development in Doha city, 1970–1997, PhD thesis, University of Salford, Salford

Al-Hahtlul S (1981) Tradition, continuity and change in the physical environment: the Arab-Muslim city. PhD dissertation: Massachusetts Institute of Technology

Al-Kholaifi MJ (2003) The traditional architecture in Qatar (in Arabic). National Council for Culture, Arts and Heritage, Doha

Al-Koni I (2012) Legends of the desert (in Arabic). Dar Al Janoub Publication

Allen DJ (2019) Lost in the transit desert: race, transit access, and suburban form. Routledge

Alraouf A (2015) Towards a holistic approach to sustainability: heritage, architecture and culture as tools for assembling identity in Qatar. QScience proceedings, Qatar green building conference 2015 – the vision, Apr 2015, vol 30. https://doi.org/10.5339/qproc.2015.qgbc.30

Alraouf A (2016) Museums as a catalyst for a new urban and cultural identity in Qatar interrogating the case of Museum of Islamic Art. In: Exell K, Wakefield S (eds) Museums in Arabia: transnational practices and regional processes. Ashgate, Abingdon, pp 151–166

Alraouf A (2018a) Interrogating Qatar's urbanity as a catalyst for building knowledge-based societies and economies. In: Noori N, Rickli J-M (eds) Strategies of knowledge transfer for economic diversification in the Arab States of the Gulf. Rasmus Gjedssø Bertelsen. Gerlach Press, Berlin, pp 53–66

Alraouf A (2018b) Knowledge-based urban development in the Middle East. IGI Global Publishing, Hershey

Alraouf A (2020) The urban megaprojects paradigm in qatar: a holistic criticism and premeditated prophecy. J Urban Res Dev (JURD) 1(1):30–44. Fall 2020. ISSN 2718-0697

Alraouf AA (2022) From spectacular to speculative gulf cities: a tale of Doha and Dubai. In: Chu C, He S (eds) The speculative city: emergent forms and norms of the built environment. University of Toronto Press, Toronto

Ateliers Jean Nouvel, National Museum of Qatar, Doha. Architectural and Museographic Design: the Desert Rose (n.d.) http://www.jeannouvel.com/en/projects/musee-national-du-qatar/. Accessed 16 July 2022

Bharadwaj V (2012) Museums, cultural jewels in the Gulf NYTimescom – Published: March 20, 2012

Black G (2005) The engaging museum: developing museums for visitor involvement. Routledge

Bolleter J (2019) Desert paradises: surveying the landscapes of Dubai's urban model. Routledge
Calleja M (2014) The philosophy of desert metaphors in Ibrahim al-Koni: the bleeding of the stone. FARAXA Publishing
Davis M (2007) Fear and money in Dubai. Topos 58:62–70
Erskine-Loftus P (2013) Reimagining museums: practice in the Arabian Peninsula. MuseumsEtc, Cambridge
Exell K (2016) Modernity and the Museum in the Arabian Peninsula. Routledge, Abingdon/New York
Exupéry AdS (2018) The little prince. Clarion Books; Anniversary edition (October 23, 2018)
Fromherz AJ (2012) Qatar: a modern history. Georgetown University Press, Washington, DC
Günel G (2019) Spaceship in the desert: energy, climate change, and urban design in Abu Dhabi. Duke University Press Books
Hooper-Greenhill E (2007) Museums and education. Routledge
Jodidio P (2020) National Museum of Qatar. Thames & Hudson
Kolo J (2016) Accidental or envisioned cities: a comparative analysis of Abu Dhabi and Dubai. In: Katodrytis G, Syed S (eds) Gulf cities as interfaces. Gulf Research Centre, Cambridge, pp 161–180
Lane RJ (2000) Jean Baudrillard. Routledge, London, p 4
Malloy R et al (2013) Design with the desert: conservation and sustainable development. CRC Press
Moolla F (2015) Desert ethics, myths of nature and novel form in the narratives of Ibrahim Al-Koni. Tydskrif vir letterkunde 52:176–196. https://doi.org/10.4314/tvl.v52i2.12
Nasser A (2015) Bedouins: the only human community wiped out by the modern state (in Arabic). Published on 29 October 2015
Ouroussoff N (2010) Building museums, and a fresh Arab identity. Published: November 26, 2010 (N.Y. Times)
Pribadi M (2014) Ibn Khaldūn's social thought on Bedouin and Ḥaḍar. Al-Jami'ah J Islam Stud 52:417. https://doi.org/10.14421/ajis.2014.522.417-433
Rabinowitz D (2020) The power of deserts: climate change, the Middle East, and the promise of a post-oil era. Stanford Briefs
Rice M (1977) National museum of Qatar, Doha. Museum Int:19.2, 78–19.3, 87. https://doi.org/10.1111/j.1468-0033.1977.tb02077.x
Rizzo A (2014) Rapid urban development and national. Cities 39:50–57
Roberts DB (2017) Qatar, securing the global ambitions of a city-state. Oxford University Press: Oxford, UK
Sims D (2015) Egypt's desert dreams: development or disaster? The American University in Cairo Press, Cairo
Qatar Satistics Authority (2021) Monthly statistics, January 2021. Available at: https://www.npc.qa/en/statistics/Statistical%20Releases/General/Qatar%20Monthly%20Statistics/QMS_PSA_85_Feb_2021.pdf. Accessed 1 Oct 2024
Wakefield S (2020) Museums of the Arabian Peninsula: historical developments and contemporary discourses. Routledge
Woodman E (2008) I.M. Pei's MIA opens in Doha, review. http://www.bdonline.co.uk/story.asp?storycode=3129204. Accessed 28 July 2022

Chapter 4
Fragments and Loading Docks: Examining the Relationship Between Urban Climate Change Policy and Design Tools in a Hotter, Dryer Climate of Melbourne, Australia

Wendy Walls

Abstract This chapter traces the evolution of recent climate policy related to heat in the inner city of Melbourne, Australia. The review highlights where institutional responses, aided by the rise of digital tools, have shifted from risk assessment towards collaborative planning models, followed by design guidelines and tools. The study also reveals the institutional layers of misalignment between government structures, policy, planning and design practice. The effect of this fragmentation in Australian climate planning and design and the privilege of science in policy development has contributed to persistent implementation gaps between research priorities, policy-led ambitions, and design in the competitive built environment industry.

Keywords Urban heat · Policy · Design · Climate change

4.1 Introduction

Australia, like much of the world, is subject to rising heat via climate change. In cities like Melbourne in the south-east of the country, the impacts of increased urban heat are well documented, including severe human health effects, loss of urban vegetation and infrastructural failure. As the projections for Melbourne suggest, in addition to overall warming, the city will experience longer and more intense droughts and heat waves, further exacerbating the damaging effects of urban heat. Despite this scenario, there are surprisingly few current examples of comprehensive heat-responsive urban design in Melbourne City. The inner-city council is well known for innovative strategic urban policy, most notably the Urban Forest Strategy

W. Walls (✉)
The University of Melbourne, Melbourne, Australia
e-mail: w.walls@unimelb.edu.au

R. Roggema (ed.), *Dry Urbanism*, Contemporary Urban Design Thinking,
https://doi.org/10.1007/978-3-031-77347-1_4

(2012a) and in 2022 became the sixth global city with appointed Chief Heat Officers, aligned with the Asht-Rock Resilient Cities program. Even with these policy innovations, there is little spatial design that anticipates the oncoming challenge of extreme heat and drought. While it is common to see small-scale urban interventions like street misters, green walls and roofs, there is a clear gap between the scales of local government ambitions and how these goals manifest in urban design. There is a need to interrogate the relationship between strategic planning and guidance and the work of landscape architects and urban designers in responding to heat and drought in warming cities, such as Melbourne.

This chapter examines policy-making as a major influence on urban and landscape design for heat in Melbourne. Through this review, the institutional and interdisciplinary challenges underpin the successfulness of implementing climate-responsive design. At the scale of Melbourne's local government, policy-making works with expertise from planning, politics, science, development, and design. However, the influence of this individual disciplinary knowledge is greatly dependent on, and framed by, institutional organisational structures, existing disciplinary relationships, and policy-making processes. Further, in a democratic system like Australia's, policymaking must negotiate with the demands and needs of the broader community, including businesses, developers, residents, and interest groups, which introduces another layer of sociological values, alongside cultural expectations, and norms. This sets up a complex and nuanced set of influential relationships on the ways that policy influences and impacts on designed outcomes.

The first part of this chapter positions the City of Melbourne (CoM) in its unique climatic condition and introduces its origins as a colonial city. This context establishes the ongoing impact of the colonial legacy through the physical traits of the city and begins to reveal where sociological and institutional values overlap in the design and planning approaches towards a climate in the contemporary urban environment. It then positions Melbourne within the Australian political system which is tiered across National, State, and local governments, recognising the influence of larger political structures on how local councils might frame or enact policy responses to climate and heat. The chapter then turns to closely examine the most influential local climate and heat policy and guidance in inner Melbourne, and how these guiding tools have been used in responding to the city's changing climate through design.

This analysis of policy and relationships reveals an evolution of institutional guidance that has been greatly influenced by political responses to extreme weather events, including major heatwaves and droughts, like the 2009 record-breaking heat and Black Saturday bushfires. However, further guiding policy development are the socio-political influences of funding models and research initiatives, including leadership roles and the arrival of data-enabled technology in environmental science research. These combined forces have shaped the transition of CoM institutional guidance from risk assessment to exploring new methods of collaborative policy-making, design guidelines and open-access tools to support designers in responding to heat. By looking across these major influences and shifts in policy approaches, the chapter uncovers how institutional relationships and power structures create a complex push and pull of policy agendas and where the design professions are just one aspect of the network of decision-makers in creating the urban climate.

4.2 Melbourne and the Challenge of Climate Planning and Design

The City of Melbourne (CoM), located on the coast of south-eastern Australia (Fig. 4.1), is well known for its unpredictable weather and seasonal fluctuations. As a cultural reference point, Melbourne is described as a city of ‘four seasons in one day’[1] because the weather changes quickly and often. The complex local weather and climate behaviours create a challenging physical space for planners and designers to work into. In addition, Melbourne is particularly vulnerable to rising heat and heat wave events, both of which are predicted to increase with climate change (B. o. M. CSIRO 2022; R. Melbourne 2019).

The physical behaviours which cause Melbourne’s variable weather are linked to wind patterns from the north and south[2] and the city’s geographic position between the Tasman Sea and the arid interior land mass. Multi-day heat waves in Melbourne occur when anticyclonic conditions, commonly known as ‘blocking highs,’ reticulate air flows over the warm interior land mass. When a blocking high lingers,[3] the circulating warm, dry air effectively traps heat over Melbourne which can produce severe multi-day heat waves with minimal night cooling.[4] The end of a heatwave in Melbourne is often marked by a ‘cool change’ when the wind shifts from a ‘hot northerly’ to a ‘cool southerly’ as the high-pressure system moves away[5] (Boschat et al. 2015; Imran et al. 2017; McBride et al. 2009; Perkins et al. 2015; Turton 2019).

---

[1] Most famously in the Crowded House song *Four Seasons in One Day*.

[2] Wind flows carry and push air temperatures from the surrounding environment, often warm winds in the summer come from the north whereas cooler air flows from across the sea in the south. This doesn’t mean that all north winds carry heat as air temperature depends on surrounding seasonal and climatic conditions. However, summer northerly winds do have a tendency to carry dry hot air along with grass pollens from the north western plains, a further contributor to urban discomfort (Jacobs et al. 2014).

[3] The immediate cause of blocking high pressure systems is unknown, however, the relationships between land and ocean surface temperatures is known to exacerbate certain factors. For instance, phases of El-Nino and El-Nina Ocean temperatures influence rainfall. When there is less rain, the dryer inland soil conditions enhance warm air temperatures, further contributing to the intensifying heat in the larger weather system(Boschat et al. 2015; Imran et al. 2017; McBride et al. 2009; Perkins et al. 2015).

[4] An extreme example occurred in November in 2017 which included 12 days above 30 °C and 14 nights over 15 °C, produced by a lingering blocking high pressure system (BOM 2017).

[5] This doesn’t mean that all northerly winds are warm—this depends on the season and surrounding climatic conditions. However, summer northerly winds do have a tendency to carry heat along with grass pollens from the north western plains, a further contributor to urban discomfort (Jacobs et al. 2014).

Fig. 4.1 The City of Melbourne is located on the southeast coast of Australia in the State of Victoria. (Google 2023)

Of course, adding to local heat behaviours and climate vulnerability is the city's physical form. The effects of materials, street alignments and built form on local environmental heat and urban microclimates are well documented. For example, this was used to great effect in the design of Masdar City in the United Arab Emirates where the streets were aligned to capture prevailing sea breezes to cool the city(Foster+Partners 2007; Lau 2012; Pfeiffer 2014; Walls et al. 2015).

However, the City of Melbourne is a settler society, initially planned with foreign eyes. The British colonists[6] lacked knowledge of the local climate, instead creating city streets and open spaces based on the European styles of the 1800s. The distinctive inner-city format, the Hoddle Grid (Fig. 4.2), was laid out in 1837, orienting from the river as the primary docking point for ships. The English surveyor, Robert Hoddle, employed the grid system to enforce orderliness and regulation in the new township (Lewis 1995). The block sizes were prescribed for convenient

[6]By 1850, 20,000 British settlers were living in the township of Melbourne, and with the Victorian gold rush, from 1851, the population boomed as over 300,000 new immigrants arrived (City of Melbourne 2022a).

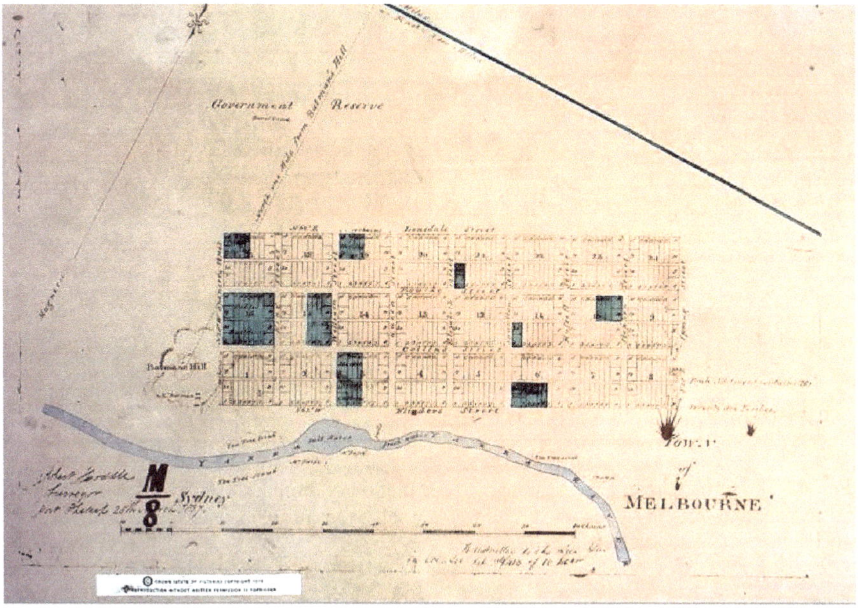

Fig. 4.2 Robert Hoddles's survey plan for the City of Melbourne shows the proposed grid system aligned to the river and main docking port. (Hoddle 1837)

subdivisions to be sold off as standard allotments; these were later cut up by lanes to provide rear access (SLoV 2023; UoM 2008).

The convenience of the grid was imposed on the landscape without attention to the physical terrain, including topography, waterways, or micro-climatic and weather patterns.[7] This led to major local environmental challenges. For example, present-day Elizabeth Street follows a natural valley in the terrain and is notorious for flooding. Likewise, the grid is set at a north-east to south-west orientation, which creates a wind channel effect converging in the inner city and creating uncomfortably high winds and turbulence at street intersections(Moya et al. 2013; Shaw 2013; Tapp 1985).

This colonial period, therefore, establishes major urban decisions that have lasting consequences for contemporary planners and designers. As illustrated in the graph from the City of Melbourne's *Urban Forest Strategy (*2012a*)* in Fig. 4.3, these colonial choices create important points of sociologically constructed attitudes towards Melbourne's planning and design. In particular, the reflection of a colonial mindset to 'taming' and 'cultivating' the landscape, where the landscape and environment are to be controlled through processes of 'reclaiming land' as part of city making.

[7] While Hoddle was aware of the need for ventilation through cities, and he proposed wide street and big blocks, this followed the 1800s miasma theory of disease not a concern for temperature control.

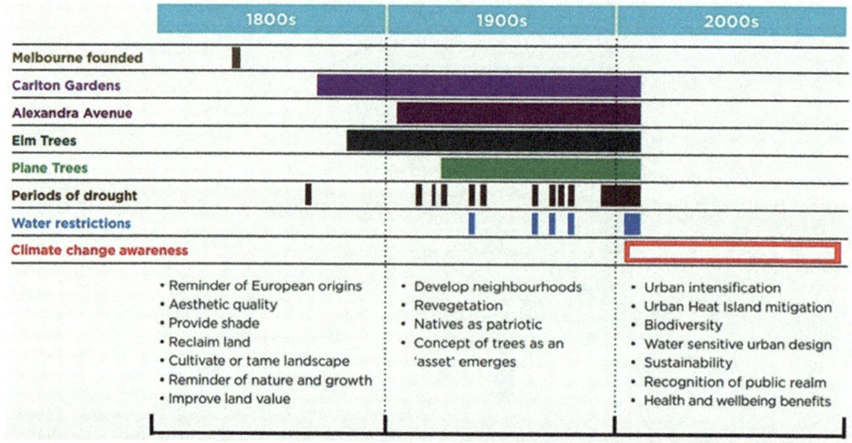

Fig. 4.3 From the Urban Forest Strategy, the evolution of attitudes and influences about urban open space and trees in Melbourne highlights key shifts towards climate change awareness in the 2000s. (City of Melbourne 2012a, p. 20)

Surprisingly this timeline presents a hard-line in the early 2000s, where climate change awareness appears to take over from earlier influences. Yet urban form cannot change so rapidly, and neither do sociological values. Contemporary urban space is unavoidably influenced by prior decision-making and embedded cultural positions. Responding to the local climate in Melbourne requires engaging with the fluxing seasons and weather, future predictions of a hotter and dryer climate, but also with the enduring colonial legacy (a sociological construct), which in some cases is now considered heritage.

This tension is clearly demonstrated in the ongoing debate over planting in the early colonial parks. Carlton, Fitzroy, Flagstaff and Treasury Gardens were all designed during the late 1800s and used introduced trees familiar to the new colonists. Most famously, avenues of Elm trees were planted, creating one of the largest and most valuable populations of Elm outside of Europe (Haining 2005; City of Melbourne 2022a, b). The suitability and resilience of the Elm trees are now a point of contention between sociologically formed heritage values and the fact that they are no longer fit for the increasingly hot and dry climate.

The dilemma over the suitability of tree species illustrates how sociological values can greatly influence a city's characteristics. Although attitudes change over time, earlier sociological perspectives such as tree preferences, can linger and continue to shape the city's physical attributes. Change does not happen overnight, and this is particularly true when looking at how the Australian government has shaped more recent climate policy in the past twenty years. While this chapter is focused on the City of Melbourne, as the following section explores, in Australia, and in many other countries, local policy and actions are greatly influenced by high levels of State and Federal governance.

4.3 Governance

The Australia government is split into tiers of Federal, State (or Territory), and local divisions.[8] The hierarchy of power and funding begins with Federal level decisions which are passed down to state and local authorities in a vertical division of authority. Within each of these tiers are discrete portfolios of responsibility and organisation.

In terms of climate policy, the institutional effect of the governance structure has historically limited cohesion in Australian climate regulation, with little agreement on climate priorities. Part of this can be explained by the vastness of the continents' different climatic and economic conditions. For example, Australian climate zones (Fig. 4.4) stretch from the tropical regions of the north down through desert, grasslands to the temperature zones of the south where Melbourne is located. While much of the Australia population live on the east coast in the major and capital cities of Melbourne, Sydney, Brisbane, Canberra, and Hobart, these are all within their own distinct climate zones and with varying climate priorities.

Further complexity emerges in the structuring of climate policy across multiple portfolios and across three tiers of governance. The fragmentation of the system

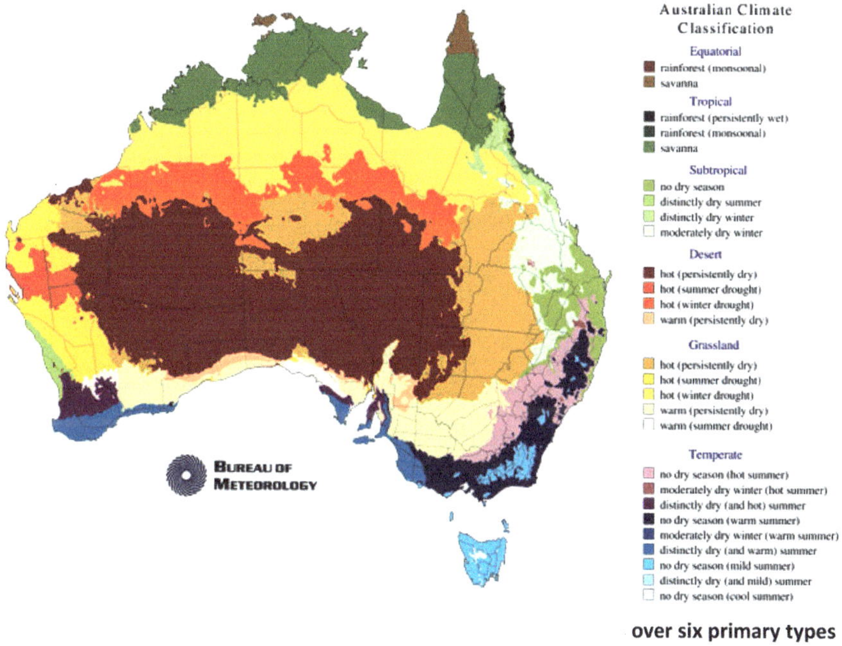

Fig. 4.4 Australian Climate Classifications, showing 27 classifications over six primary types. (Bureau of Meteorology 2023)

[8] All of which exists under sovereign rule as a constitutional monarchy as a British colony.

creates isolation between decision making bodies, duplication, redundancy and even opposing approaches to the same sets of issues. This is highlighted in the Low Carbon Living discussion paper on *Cooling Australian Cities (*2017a, *p. 34)* which states;

> Despite, or perhaps because of, successive policy reforms, the urban planning and environmental governance system in Australia remains fragmented and complex. There is a lack of portfolio alignment between energy, climate change and urban planning institutions between and among the jurisdictions. Such policy governance creates confusion, overlapping of issues and, at times, tensions between institutions and jurisdictions.

Australia is divided into eight states and territories. The small State of Victoria alone has 79 local councils operating largely independently from one another.[9] Thirty-two of these are within the metropolitan area of Greater Melbourne where the City of Melbourne represents one council area specific to the inner city (Fig. 4.5).

The fragmentation and division of authority begins at the Federal level which accounts for shaping national priorities on climate but also negotiates with global agreements. For example, Australian climate policy on carbon emission reductions is influenced by international targets like the United Nations Climate Change agreements. Similarly, the Toronto (1990), Kyoto (1997)[10] and Paris (2016)[11] agreements have all affected Australian greenhouse gas emissions targets.[12]

In the Federal government structure climate policy generally sits within portfolios of the environment, energy, water, or resources. However, these portfolio structures are often reshuffled, particularly when there are changes to government and

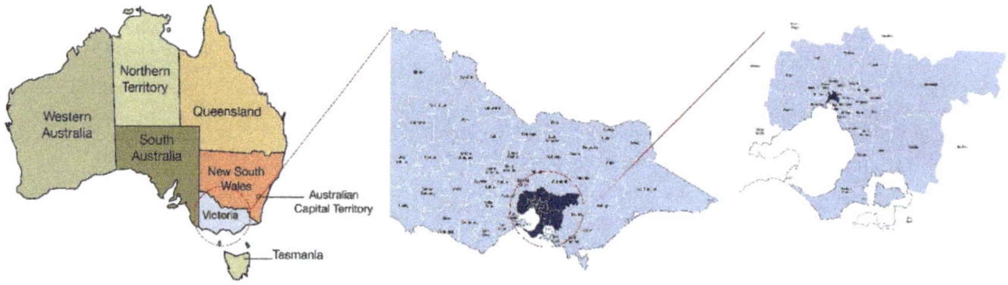

Fig. 4.5 The state of Victoria within Australia is divided into 79 local council with 32 of these are within greater metropolitan Melbourne which covers an area of approximately 9900 km^2 with over five million people. The central inner council City of Melbourne is 37 km^2 with a resident voting population of 169,860. (ABS 2021)

[9] Each local council has an elected Mayor and supporting councillors (Local council elections are held every four years.) who represent immediate community interests. The borders and size of local electorates are determined by population to ensure equal voter representation (VEC 2022).

[10] Initially, to reduce greenhouse gas emissions 5% below 2000 levels by 2020.

[11] To reduce greenhouse gas emissions by 26–28% below 2005 levels by 2030.

[12] Or in the case of the 1997 Kyoto agreement when Australia, under John Howards Liberal government, initially rejected the treaty. While the Kyoto protocol was eventually signed in 1998, Australia still did not ratify until 2007 under a Labor Government (MacNeil 2021).

switching between the primary parties of Labor and Liberal Coalition which frequently happened through the 2000s. At the time of writing, under a federal Labor government, the four areas are combined into a single portfolio—*Climate Change, Energy, the Environment and Water*. However, this doesn't mean that all areas related to climate governance are also held within the same portfolio. The related areas of climate science, research and planning are distributed across portfolios including Industry, Infrastructure, and the environment.[13] Within each portfolio divisions are discrete department agencies, government appointed boards and other advisory committees. For example, Fig. 4.6 shows under the *Climate Change, Energy, the Environment and Water* portfolio, is the National Bureau of Meteorology and Climate Change Authority. Whereas the *Industry, Science and Resources* houses the major science institutes including CSIRO, the Australian Cooperative Research Committee and the building ministers and Australian building codes boards. These are again separate from the Cities Reference group, Infrastructure Australia and Creative Partnerships which are under the *Infrastructure, Transport, Regional Development, Communications, and the Arts* portfolio.

This structuring and division offers a useful illustration of how political issues and social values are understood in relationship to one another at the institutional

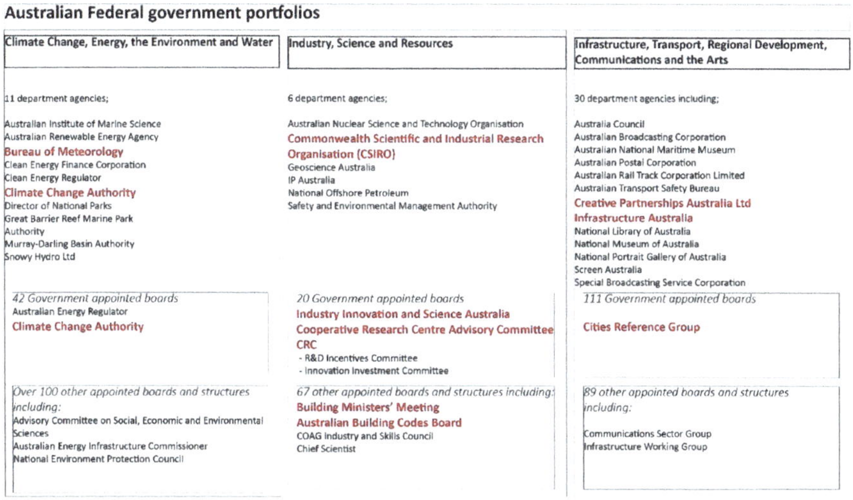

Fig. 4.6 Climate research, science and urban policy are divided across three of the 15 Federal portfolios

[13] The 15 Australian Federal Government portfolios include Prime Minister and Cabinet; Treasury; Attorney-General; Home Affairs; Employment & Workplace relations; Health & aged care; Social Services; Education, Finance, Foreign affairs, and trade; Defence & Veteran affairs; Agriculture, fisheries & forestry; Climate, change, Energy, the environment & water; Industry, science, and resources; infrastructure, transport, regional development, communications, and the arts.

scale. For example, there are no science organisations in the infrastructure and development portfolio. Likewise, the environmental scientific research agencies are separated from industrial research, all of which are separate from creative practices. Thus, beginning from this high level of Federal governance, the organization of the institution acts to shape disciplinary alignments and allegiances which have a major impact on how these forces work on the ground in designing for urban heat and drought.

This fracturing across institutional organisations and division of agendas is repeated at State and local council levels. While much of the negotiation and agreement on overarching climate and environmental priorities occurs at the National level, actions are delivered through State and local planning. For example, State and Territory governments are in control of large-scale planning, public health, education, roads, and land use.

In the case of Victoria, there are nine State portfolios and agencies under the 2023 state Labor government. As with the Federal system, urban development, climate planning and design issues are dispersed across these divisions. For example, the office of Victorian Architect sits under the *Premier and Cabinet* portfolio, whereas Urban planning and Open Spaces reside with the *Transport and Planning* portfolio.[14] These are separate again from Climate Action and from emergency services or the heat health advice—all of which contribute towards establishing responses to managing climate effects of heat in urban environments.

At the Victorian state level, these divisions also begin to reveal replication between issues and responses. For example, the two major current programs focused on urban cooling exist in separate agencies. *Cooling The West*, a tree planting program for Melbourne's western suburbs, is part of the Department of *Energy, Environment and Climate Action* (DEECA 2023). While the *Cooling and Greening Melbourne* strategic plan[15] for increasing urban vegetation is run by the Department for Transport and Planning (DTP 2023a).

While these cooling programs and strategies are connected under the State government, the layering of responsibilities begins to expose the complications of vertical distributions of authority with local councils and how these impact funding and project implementation. For example, *Cooling the West* is a grants program where local councils can apply for State funding for tree planting and where the uptake is dependent on the priorities of specific local councils.

The same story of fragmentation plays out again at Local council level which is responsible for town planning, provision of services like water and waste and community recreational facilities (ABC 2022). Local councils make specific decisions about land-use planning, infrastructure provision and major projects which takes on the addition of operational responsibilities (DCCEEW 2022; Hanna 2022). At this scale, portfolios are distributed to elected councillors in addition to an executive

---

[14] In January 2023, the Department of Transport was renamed to the Department of Transport and Planning (DTP) Previously, Planning and Transport were two separate portfolios (DTP 2023a, b).

[15] Part of *Plan Melbourne 2017–2050* (2017b).

organisation, adding further divisions to the structuring and alignments of departments and operations. Figure 4.7 shows these two parallel structures for City of Melbourne. The first structure comprises the 13 Councillor portfolios, while the second is the executive structure made up of five primary divisions. Again, climate planning urban development and design are spread between portfolios of *Environment* and *City Planning* but there is the addition of specific *Sustainable Building* portfolio which incorporates building design and green roofs. This is separate again from Sustainable Development goals and benchmarking, which is held in *Finance, Governance and Risk*. Even more confusingly, these are divided again in the executive organisational structure. Here Climate Change, City Resilience, Parks, and Greening are linked to *Strategic Planning and Climate Change*. However, City Design, projects and infrastructure are run by *Property, Infrastructure and Design*.

The three structures of Federal, State, and local portfolios highlight the considerable challenges within the Australian government structure to align interests, research, and funding on climate change issues. The separation of powers occurs vertically through the three-tiered system and again horizontally within the organization of each layer of governance.

However, even further complicating this governance is the polarising nature of Climate policy in Australia where culture and social attitudes and worldviews towards climate and environmental issues are influenced by histories and lineages of prior decision-making (Brown and Havstad 2017; MacNeil 2021, p. 164). There is extensive scholarly work in discussing how Australian attitudes toward the environment, weather and climate are tied to the construction of colonial Australian cultural identity (Ferng et al. 2020; Ireland 2003; Tranter and Donoghue 2007). Popularised colonial Australian characterisations, from the battler to bushman, are

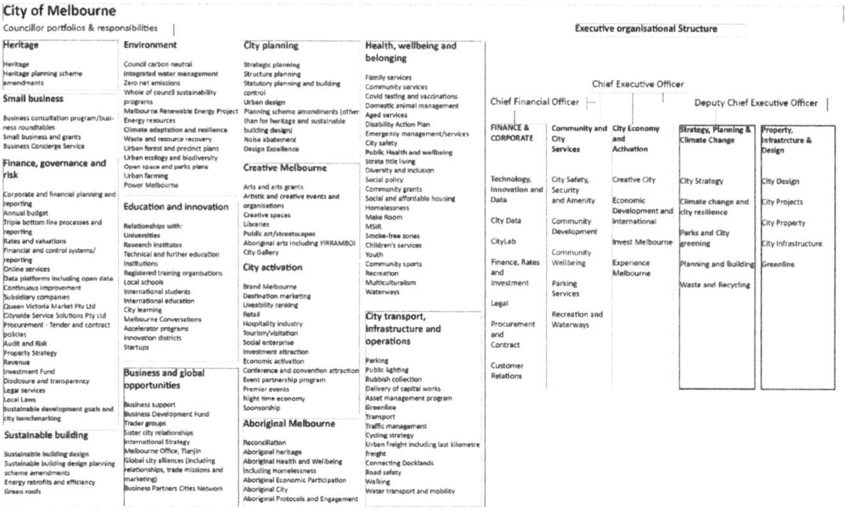

Fig. 4.7 The inner-city council of City of Melbourne is organised into councillor portfolios and an executive office. Climate planning and design responsibilities are divided across the structure

bound up with living in and surviving the harsh Australian climate. Natural disasters, including drought, heatwaves and fire, are considered obstacles to be overcome through hard work and endurance (Ireland 2003, p. 61).

The background cultural narrative is one of a hostile Australian environment, a place of extremes. For some, the challenges of incremental warming, such as the two degrees Celsius in climate change projections, are of minimal concern compared to the narrative of extremes. As Banwell, Dixon, Bambrick, Edwards, and Kjellström (2012, p. 3) reflect, 'It [heat] is accepted as part of the Australian way of life and integral to the habitus.' So even with growing concern and progressively agreed-upon scientific evidence of worsening climate change impacts, there are residual, yet enduring social and political attitudes towards the experience of heat that downplay the risks.

Australian politics reflects these entrenched sociological values through a conservatism and general wariness of climate-based action(Crowley 2021; Dubash 2021). Since Australia first adopted the Toronto greenhouse emission reduction targets[16] in 1990, Federal-level climate action has struggled to gain momentum(MacNeil 2021). The lack of climate action was exacerbated into the 2000s and 2010s by a frequently changing elected national government. During this period, the frequent switching between the Liberal Coalition and Labor governments with internal leadership challenges, led to substantial portfolio reshuffling. Since 2007 the responsibility of climate change issues has moved from its own portfolio to over ten versions of alignments with energy, water, environment, sustainability, agriculture, and emissions, further exacerbating the fragmentation and complexity of climate governance described above.

The turnover between governments and portfolios, and the shifting attention to and from climate change issues over this period has directly impacted resourcing and action. For example, in 2007, the newly elected Labor government ratified the Kyoto agreement, which the Liberal coalition had previously rejected. Between 2007 and 2013 the Labor government established new reviews and committees, including the Multi-party Committee on climate change and the Climate Commission, which MacNeil (2021, p. 170) describes as a high-point in Australian climate action. However, much of this work was overturned or decommissioned following the 2013 election of the Liberal coalition.

The back and forth between national leadership and funding models has created an inconsistent approach to managing climate change. Not only is there a vertical diffusion of authority that is subject to political cycles (Dubash 2021, p. 56) but there are also rifts between the policies, institutions, research and major stakeholders. MacNeil (2021, p. 164) summarises:

> Institutional underdevelopment has hindered Australian climate policy for the better part of three decades, leaving policy efforts sporadic, generally unambitious, and highly vulnerable to political attack and rollback.

---

[16] Much of Australian climate policy and debate is focused on electricity production and use as the driving narrative climate change at the very large scale of global warming. In part this reflects an international discussion on emissions and temperature reduction targets.

This brief outline and background to the Australian governance structure highlights the messiness and tensions that are embedded in the organisation of policy, funding models, and research. In terms of planning and guidance on climate change and urban heat, there are few simple or easy alignments cutting through the tiers of government where political volatility has hampered continuities. Forino, von Meding, Brewer, and Gajendran (2014, p. 481) explain:

> Australia has a long history of fragmenting decision-making powers which set a precedent in planning policy where an inability to find a common agenda is an acceptable norm.

Despite continued consideration of climate, very little policy has maintained momentum over time without interruption of political cycles and policy backflips which have been common to the Australian climate response. The following section shifts scale to examine the City of Melbourne´s local policy more closely. While Federal and State governments influence major funding and Australian climate priorities, it is municipal councils who most often deliver actions. At the local level, urban planning, building regulations and the design of streets, parks and open spaces are the primary mechanisms where institutional goals and ambitions intersect with the spatial and physical characteristics of sites.

4.4 Planning in the Inner City of Melbourne

The City of Melbourne's (CoM) first *Climate Adaptation Policy* was published in 2009. While there was significant earlier environmental planning, including water sensitive urban design research, the late 2000s marked the beginning of local policy efforts focused on specific climate change issues such as urban heat and drought. This moment is recognised as a major turning point towards climate change awareness, as already shown in the earlier City of Melbourne diagram (Fig. 4.3). This section maps out the evolution of climate research and policy from 2009 to 2015, looking for where and how heat is addressed, before turning to examine the most recent set of urban design guidance and tools. Through close analysis and comparison of these primary responses, this analysis reveals the environmental, social, and political catalysts which have influenced the direction of local policymaking and planning for the challenges of urban heat. It becomes clear that rather than proactive government agendas at the national or state level, the significant climatic challenges of drought and heatwaves felt directly in Melbourne during the 2000s are the first major catalyst for policy change.

4.4.1 2009–2011: Risk Assessment and Emergency Response

The period from 2009 to 2011 can be considered a pivotal moment, where CoM first becomes concerned with developing planning policy to guide urban heat responses. While innovative, these early policy moves were framed by the need to create a comprehensive argument for why heat might be a serious urban concern. For example, the first *City of Melbourne Climate Change Adaptation Strategy (*2009*)* was developed as a risk assessment plan. The document weighs the *potential* threats of climate change as a set of 'cascading consequences' and acknowledges the connections between environmental and human systems (2009, p. 7). Considered a milestone in local planning for acknowledging the long-term challenges of the changing climate, the strategy became a key instigator in the conversation about city-led mitigation actions.

Produced at a time of rare political alignment of Labor government at Federal and State levels, the *City of Melbourne Climate Change Adaptation Strategy (*2009*)* resulted from an inter-governmental collaboration of national and local agencies. As discussed earlier in this chapter, 2007–2013 was a high-point in Federal level climate action in Australia, which is reflected in the establishment and funding of the *National Climate Change Adaptation Research Facility (NCCARF)* (2008–2019), the *Victorian Centre for Climate Change Adaptation Research* (VCCCAR) (2009–2014) and the *Local Adaptation Pathways Program (*DCC 2009, p. 2). The CoM Adaptation Strategy was partly funded by the Local Adaptation Pathways Program—in itself, a significant early step in acknowledging that large scale (national and international) institutional guidance for climate change is most often implemented through local councils and communities (Brennan et al. 2011).

With a focus on the City of Melbourne, the Strategy identifies *heatwaves* and *extreme irregular rainfall* as the two most significant weather events impacting the inner city and most likely to be exacerbated by climate change (2009, p. 2). While the document points to the then-established body of Water Sensitive Urban Design (WSUD)[17] research for storm water management, at the time, there were no explicit design or control measures for managing heat or severe lack of water at the city scale. In fact, the adaptation policy makes few recommendations for managing urban heat, instead reflecting cultural expectations that heat is standard to the Australian climate, noting that Australians have a 'propensity to participate in events in very hot conditions' (DCC 2009, p. 80). Here the strategy provides a direct example of where the sociological values related to colonial attitudes of endurance and hardship discussed earlier have shaped the Australian institutional response to climate and heat.

[17] A considerable body of research in Water Sensitive Urban Design (WSUD) was produced through the 1990s and 2000s, some of which foreshadowed the influence of water for cooling the city; largely through vegetation health and soil moisture but also the potential for contributing towards localised microclimate effects (Coutts et al. 2013).

This attitude surprisingly prevailed, even though Melbourne had just lived through the 1996–2010 *millennium drought*, which created severe water shortages and large-scale environmental damage and forced a comprehensive review of Victorian water supply (DoE 2016). Even as the dangers of the urban heat island (UHI), major heatwaves and drought were progressively flagged by researchers throughout the 2000s, (Coutts et al. 2007; CSIRO 2007; Nicholls et al. 2007; Pittock 2005), there remained a political reluctance to mandate on heat and local climate effects (Heberger 2012, p. 114).[18] The major contribution of the initial *Climate Change Adaptation Strategy* was a series of recommendations for further data gathering and research to establish the *minimum responses* which could balance between expenditure and action—the document (2009, p. 14) explains:

> The political feasibility of adaptation actions or the cost of these being delivered can also challenge the success of adaptation plans… the aim is managing risks to accepted levels of tolerance.

However, in this same year, the unprecedented 2009 January *Black Saturday* heatwave[19] and accompanying catastrophic bushfires across Victoria would become the catalyst for an even greater policy rethink. The extent of heat-related deaths, hospital admissions, infrastructural collapse, and economic burden far outweighed any of the previous predictions of the effects of extreme heat and long-term drought. There were 374 excess deaths and 46% increase in ambulance call-outs. The total economic cost of the 2009 heatwave is estimated at $800 million (City of Melbourne 2017a).

The crucial lessons from the 2009 heatwave illustrated the immediate danger of the warming climate. This shifted the discourse on heat from being a potential future threat towards provoking new political interest in action and the beginnings of significant research funding towards understanding and mitigating Melbourne's urban heat. The resulting governmental review of the 2009 events included recommendations for improved emergency response and heat health warning systems (D.o.H.S 2009). However, to trigger an emergency response would require defining exactly what constitutes dangerous heat conditions.

These recommendations and the need for a more precise definition of heat threats and vulnerabilities spurred a next wave of local urban climate research (Loughnan et al. 2010a, b). For example, heat vulnerability mapping combined demographic, spatial and environmental data to reveal the suburbs and populations at greatest risk in heat events (Loughnan et al. 2009). This model of Melbourne introduced a new level of detail about the relationships between the city, socioeconomics, and environmental risk, using heat to reveal spatial inequalities in the urban form.

---

[18] Heberger (2012, p. 116) describes the response to the Millennium Drought and the rhetoric of some politicians, who describe restrictions by water suppliers as "draconian impositions on individual freedoms" as politicians who are reluctant to 'disrupt the status quo in the absence of a strong demand from the electorate.'

[19] The extended heat throughout days and nights during the January 2009 heatwave was the first time that Melbourne's daily mean temperature exceeded 35 °C, significantly above the 30 °C threshold (BoM 2009).

New policy quickly followed, including Melbourne's first *Heatwave Response Plan* (2010). The strategy focused on awareness to heat events, including aligning temperature thresholds[20] with emergency services and improving communications. Further establishing a local heat health alert system and a heatwave evacuation plan for Melbourne's CBD with locations for sheltering 297,000 inner city workers (Dow 2015, p. 13).[21] While these measures reflected increased responsiveness to immediate threats, the approaches also highlighted problems with existing plans. For example, the primary message when a heat health alert is issued is to stay indoors, preferably in air-conditioned space.[22] But as the 2009 heatwave had demonstrated, extreme heat can disable power supplies and without the aid of air conditioning, most of Melbourne's building stock becomes dangerously uninhabitable (Trounson 2017).

There were still unanswered questions about how to better prepare the city for long term temperature change, and drought, particularly in the external environment of Melbourne's streets and outdoor spaces. Although awareness of the dangers of urban heat were increasing, there was relatively little knowledge of local outdoor heat mitigation strategies or the effectiveness of urban design methods for creating cooling effects. For example, a 2013 study of Melbourne based built environment professionals including architects, landscape architects and urban designers found that very few understood the urban heat island effect or localized temperature variance and few perceived that design had little influence over outdoor urban temperatures (Trundle et al. 2013, p. 34).

This early period in Melbourne's climate planning marks an important shift in how the sociological perspectives on climate move from acknowledging potential climate risks to the far more challenging question of managing them. While not yet formalised within policy documents, the move from risk to action is a significant turning point in planning, which also moves closer to what happens in design. This need to act forces decision-making in a way that had not been previously grappled with. In the next major phase of policy development, attention turned to identifying the supporting evidence from environmental science research that might help with action-based decisions.

---

[20] The Victorian Department of Health (2015) defines this as an average 24 hour temperature of 30 °C or more. Above this range, 'heat-related illness and mortality increase substantially.' For example, if the maximum daily temperature is 40 °C, and the overnight forecast temperature is 20 °C or more, this triggers a *heat health alert* to service providers and emergency response in anticipation of increased heat-related illness.

[21] This was specifically for scenarios where power fails to run office air conditioners and lifts.

[22] This is still the key message for heat wave response - the current heatwave advice from the City of Melbourne (2018a) suggests 'If it is too hot where you are living, consider going to an air-conditioned shopping centre, library or movie theatre. Or visit a friend or family member that has air conditioning in their home.'

4.4.2 2012–2015: Building on Scientific Evidence for Decision-Making

The period of 2012–2015 highlights the quickly growing interest in managing urban heat. As local scientific research produced increasingly detailed information about Melbourne's climate effects on the city, policy led guidance followed with a growing diversity of strategy, guidelines, and support. To properly understand this transition from climate assessment to action, it is crucial to recognise the influence of data and simulation technologies that emerge in the mid-2000s.[23] Data-capture technologies like LIDAR mapping, thermal imaging, and environmental sensors had become increasingly affordable and available through the 2000s. For the City of Melbourne, the new data sets and tools were critical to building evidence of the impacts of heat and informing a new generation of planning responses.

The greater range of data types offered a clear picture of local climate effects. For example, the thermal image from the *Urban Forest Strategy: Making a Great City Greener 2012–2032* (2012a) shows urban surface temperatures (Fig. 4.8). This stationary mapping taken by satellite shows the accumulation of heat in hard surfaces (red and orange) and the resulting temperature difference across the during night-time, offering visual proof of the urban heat island effect alongside the relatively coolness (blue) of open and vegetated areas.

In this case, the data-enabled technology of satellite thermal sensing provided a critical layer of evidence, or proof of local climate effects like heat to support arguments for policy-based climate action. The collaborative work of environmental science and engineering produced much of this kind of early data-driven analysis in Melbourne, including measuring external materials and surfaces to quantify energy efficiencies in building envelopes. For example, 2012 the Victorian Centre for Climate Change Adaptation review of Green Infrastructure explicitly focused on 'quantifying the cooling and energy-saving benefits' (p.iii) of plants grown outside and on buildings. Similar research included *City of Melbourne Cool Roofs Guide* (2012c), which tested the effect of lighter roof colours by recording the change in internal room temperatures and further calculate the reduction in internal energy use.[24]

[23] In this period, the fruition of the research centres established in the previous period began to push out policy support tools, including *Supporting evidence-based adaptation decision-making in Victoria* (AECOM 2013) and *Policy Guidance Brief for Managing Heatwaves* (NCCARF 2013, *p. 5*). This guidance focused on how to align scientific research and provide measurable benefits. As the NCCARF *Policy Guidance Brief for Managing Heatwaves (2013, p.5)* states, 'public open spaces should be redeveloped alongside extensive monitoring of conditions from a proper research perspective, in order to develop a 'strong, well-coordinated evidence base.'

[24] This research also tested several paint materials, building types and locations, further comparing field results with digital simulations. The final report found that by raising the roof surface albedo (painting it white) and reflecting solar radiation, the internal temperature of the building could be significantly (up to 18 °C) reduced (p. 41).

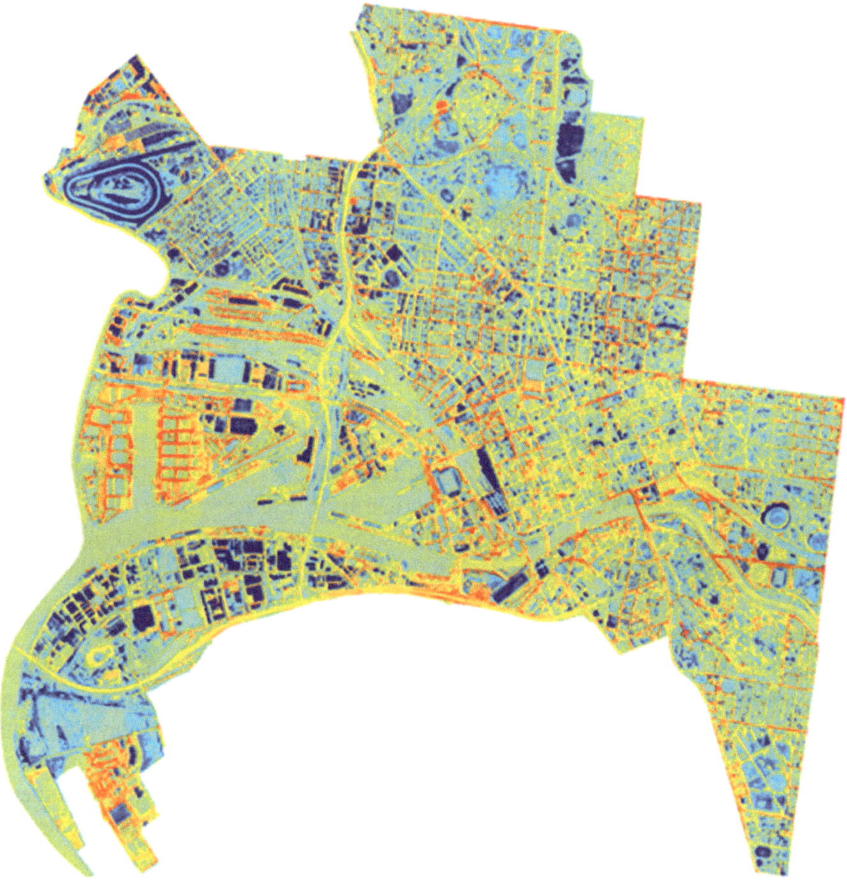

Fig. 4.8 Thermal image of inner Melbourne city. (City of Melbourne (Cartographer), 2012)

As attention turned more towards the outdoor environment and larger scale thermal mapping, like Fig. 4.8, the data-driven practices of measuring and quantifying were extended as the primary method of bridging environmental science, with the social and economic interests of the built environment. This is illustrated in one of Melbourne's most influential policies focused on external open space—the *Urban Forest Strategy: Making a Great City Greener 2012–2032* (2012a) which was a landmark project[25] for large-scale urban planning.[26] Notably, this was one of the

[25] In 2014, City of Melbourne won an international 'City Climate Leadership Award' through the C40 cities network for its urban landscapes' adaptation program and research support.

[26] Developed in relationship with the Australian Research Centre for Urban Ecology (ARCUE) and CSIRO Ecosystem Sciences, the policy developed from existing policies and global best practice.

earliest examples of Australian planning for wholistic[27] notions of the external urban environment.

While not exclusively addressing urban heat or climate, the *Urban Forest Strategy* makes explicit connections between the value of trees for air quality, habitat and temperature reduction and the sociological values of the economy, community, and human health. Further, the urban forest strategy acknowledges that most Australian infrastructural and design decisions are based on economic cost benefit. The strategy expands work with data to financially substantiate the many potential gains of urban trees.[28]

In doing so, the strategy emphasises the legitimacy of the research and highlights connections to scientific authority (2012a, p. 22) through statements like:

> The City of Melbourne has prepared a scientifically based formula for calculating the amenity value of our trees. The formula is based on factors including tree condition, species type and growth rate, aesthetic values, and locality values.

The benefits of urban trees were calculated using *i-Tree Eco,*[29]a data-driven tool[30] which calculates the amenity of existing trees such as carbon sequestration, air and water quality, and energy savings through shading buildings into a dollar amount. The calculations translated the environmental and social benefits of one tree and then multiplied over a whole tree population. At the time, the estimated amenity of Melbourne's whole urban forest was $AU700 million (2012a, p. 14).

The results provided a simple message, the more trees, the better. By presenting a rationalised economic argument, based on data and science, for the snowballing benefits of urban greening, the strategy was ground-breaking in justifying urban vegetation and prioritising the external urban environment. Importantly, the simple message of needing more trees allowed the scheme to set clear targets for the whole city region, including the 40% canopy cover by 2040 (2012a, p. 7) which remains in place today.

In support of the tree canopy targets, the strategy also identified the need for more *Australian specific formal guidance and standards* to help designers and planners integrate vegetation into the city. These calls for more *evidence-based tools* and *technical guidance* for design were reinforced by parallel research from national research projects like the Cooperative Research Centre for Low Carbon Living's (CRCLCL) Urban Micro-Climate research project which had also highlighted the

---

[27] Concurrent to the release of the *Urban Forest Strategy* was the City of Melbourne's *Open Space Strategy: Planning for Future Growth* (2012b). The core ambition of this plan was to provide more open spaces in walking distance of residential areas for community and environmental benefits including managing the urban heat island through the integration of greening.

[28] Healthy urban trees are positioned as the core solution to resolving a range of impending climatic threats, from less water to urban heat and the increasingly dense urbanised environment.

[29] i-Tree Eco was developed by the USDA Forest Service, Davey Tree Expert Company, The Arbor Day Foundation, Society of Municipal Arborists, International Society of Arboriculture, Casey Trees, and SUNY College of Environmental Science and Forestry (2006).

[30] The tool also simulating management strategies and future effects on tree population.

missing link between 'scientific research and the planning/design professions' (LCL 2017a).

Bolstered by success of the *Urban Forest Strategy*, Green Infrastructure (GI) quickly dominated Melbourne' approaches to environmental issues in external space, including urban heat. Following this momentum came a series of government-funded research projects specifically aimed at technical guidelines for growing plants in cities. Outcomes from this period include VCCCAR's *Green Infrastructure Guide (2013)*, and the *Growing Green Guide: A Guide to Green Roofs, Walls & Facades in Melbourne and Victoria, Australia* (DEPI 2014). Through this work, Melbourne garnered international attention as a leader in urban climate adaptation. In 2014, the City of Melbourne won the 'City Climate Leadership Award' through the C40 cities network for its urban landscapes' adaptation program and research support.

However, once again, the evolution of Melbourne's policy responses to heat was re-directed by a major environmental event. In early 2014, the City of Melbourne experienced another record-breaking heat wave. Extreme high day temperatures above 41 °C with minimal night time cooling extended over four days, becoming Melbourne's longest and most intense heatwave on record (D.o.H 2014). Once more, the impact on human health was severe as emergency and health services struggled to respond to demand.

While the earlier *Heat Wave Response Plan* (2010) had signalled these potential pressures on the system, and offered practical actions, the impact of the 2014 heatwave revealed the gap in communicating that information to the wider community. The 2014 state-wide Victorian Auditor-General report *Heatwave Management: Reducing the Risk to Public Health* (2014) found that the activation plans were poorly communicated, did not target the whole community and were not well understood (p. 14).[31] This was particularly evident in vulnerable communities such as elderly who have minimal media engagement or in migrant populations where English is a second language. The report concluded that responses to heatwaves were being managed at an 'operational rather than a strategic level' (p. 10). Further recommending a focus on equipping the wider community with multiple options for coping with heat, including the provision and access to equitable resources and cooler locations.

Thus, political focus on urban heat swivelled back to community responsiveness and action-based emergency- type initiatives for building communications and community resources. For example, the City of Melbourne *Cool Places* map (2015) shows cooler public locations (Fig. 4.9) within inner Melbourne city. The more recently developed *Heatwaves and Homelessness Action Plan* (2018b) offers swimming pool passes and free movie tickets as respite from heat waves. While these provide important and tangible community-based actions, the questions of equity

[31] The report also comments on unclear governance arrangements and variable quality of planning across the state, for example, 'Victoria has a state-level emergency plan for tsunamis, it does not have a state-level plan for heatwave's (VAGO, 2014, p. vii).

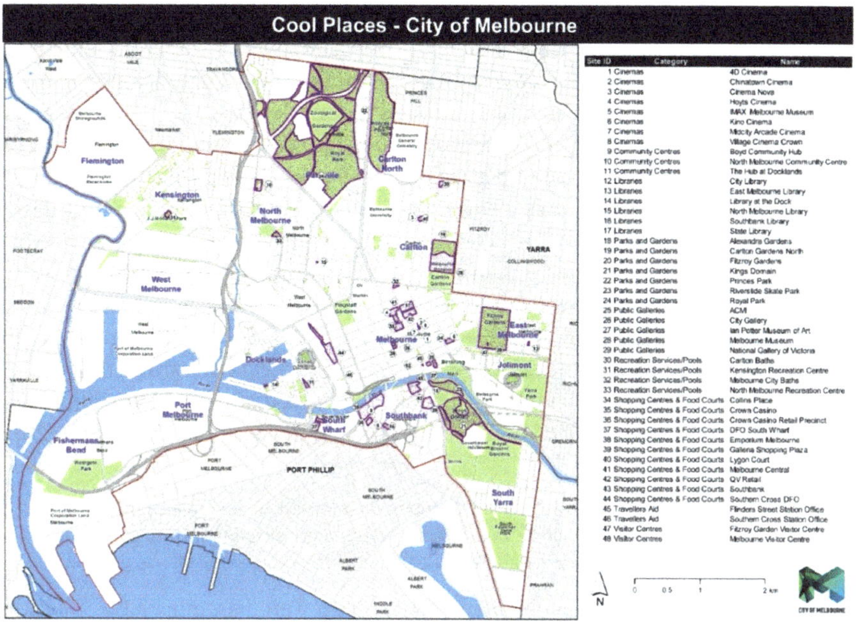

Fig. 4.9 City of Melbourne cool places map and list of locations offering public amenity cooling including shopping centers, parks, swimming pools and cinemas (City of Melbourne 2015)

raised through the Auditor General report articulated more nuanced issues regarding urban structures, long-term design and demographic vulnerabilities.

Much of the existing guidance and policy made assumptions about private amenity and access to resources. For example, air conditioning is still considered the safest and most effective approach to keeping people cool. From the list of 48 cool locations in the *Cool Places* map (Fig. 4.9) only seven are parks and gardens and the remaining 41 are indoor shopping centres, cinemas, and community services that rely on air conditioning. However, for many, rising electricity costs make running air conditioning prohibitive.[32] For example, elderly Victorians reported restricting the use of home air conditioners to conserve power during heatwaves (VAGO 2014, p. 39). Without the mediation of air-conditioning,[33] many internal spaces become dangerously hot.[34]

Over this period of policy development from the initial risk assessment approaches to emergency response and building evidence for the value of trees in

[32] Correspondingly, accessing air-conditioned public spaces requires travelling which presents specific challenges of safely moving through urban space in very hot conditions.

[33] The energy used by air-conditioning units during heatwaves contributes significantly to peak power demands and the likelihood of power outages (Cornes and Cook 2018; Saman et al. 2014).

[34] As a later 2016 study of Melbourne apartments tested six apartment designs in the City of Melbourne. All of them 'failed heat stress tests, exceeding international health standards should air-conditioning fail during a heatwave' (Thrive 2017).

cities, the scale, variability, and intensity of environmental heat events were major drivers in revealing the systemic complexity of planning and design for Melbourne's fluctuating climate. From the cost of electricity to poorly insulated homes and lack of access to amenable public spaces through to the environmental performance of urban external space in highly variable climate conditions, the effects of heat are multi-scaled and shift between long term adaptation plans and short-term actions.

Managing this considerable complexity required a shift in policy and design approaches. Flagged by the 2014 Auditor General report, planning for heat events was dominated by operational rather than strategic responses. Even though the Urban *Forest Strategy* presented long term plans like the 40% urban canopy cover target, this would take years to eventuate, with a slow build towards impacting the local climate. Regardless of the justifiable and proven benefits of urban greening, there were still gaps between future planning and immediate urban adaptation measures. Reflecting on this time, Forino, von Meding, Brewer, and Gajendran (2014, p. 474) state:

> the trend in climate change adaptation was focused on top-down governance measures disseminated from an environmental science perspective, and involving information distribution, awareness and education programs.

For built environment designers, there were still no well disseminated methods for working with the fluctuating climate. The comprehensive technical guidance on Green Infrastructures, linked to data, simulation, and the rigour of scientific approaches, all contributed to a perception that urban trees were the only option when designing for urban heat. However, the deceptively simple approach of adding more trees soon ran into the challenges of working in complex urban sites, such as limited access to soil, service provisions, water restrictions and community specific demands. Even though the *Urban Forest Strategy* had positioned trees as wholistic solutions to multiple issues like heat, air quality, health effects and economic benefits—the plan was also restricted to a single council region withing the much larger metropolitan area and the outcome was focused on the single solution of trees.

In addition, the shifting focus between long- and short-term plans also highlighted the major challenges of diverse community needs in both spatial and socio-economic terms. There was an obvious need to better understand contextual barriers to climate planning but also to connect urban climate strategy with concurrent policy on water management and open spaces. This emerging interest in more holistic planning was catapulted into the next phase of Melbourne's' climate planning and policy when the City of Melbourne was accepted into the international 100 Resilient Cities program in 2015.

The influence of the 100 resilient cities program included funding new leadership positions in city government, alongside logistical guidance towards creating *holistic Resilience Strategies* across the Greater Melbourne region (Kent 2016; R. Melbourne 2016; Rockefeller 2013). This desire to connect across multiple municipalities and existing policy challenged the much of the existing institutionalised fragmentation of political powers and municipal divisions outlined earlier in this chapter. In a major structural change, climate policy was no longer produced from within the isolated silos of local councils or from single issue perspectives. Further, by forcing the connections between climate issues across geographic regions, local policy making necessarily expands towards exploring new multidisciplinary policy making processes. This instigates new models of collaboration between policy makers, urban ecologists, developers, planners, and designers. This in turn, leads to a more diverse set of approaches to responding to urban heat.

4.4.3 2016—Present: Resilience and Collaborative Planning

From 2016 onwards, institutional guidance in Melbourne has shifted from largely analysis-based environmental science towards a more diversified set of long- and short-term planning strategies. The final section of this chapter focuses on this more recent period of policy making looking specifically at how the emergence of collaborative planning processes support multidisciplinary efforts in built environment design.

The 100 Resilient cities (100RC) program was a co-funding model established by the North American philanthropic[35] Rockefeller Foundation. The program provides external support including financial resources, but also requires a structural and economic commitment from the partner city. For example the City of Melbourne committed to at least ten per cent of the city's annual budget to resilience building initiatives, co-funding a Chair of Resilient Cities as a shared role between council and the University of Melbourne, and relating existing Council plans across the greater region of Melbourne's metropolitan councils towards creating *holistic Resilience Strategies* (Kent 2016; R. Melbourne 2016; Rockefeller 2013).[36]

The inaugural Chair of Resilience, appointed in 2016, was Professor Lars Coenen. With a background in economic geography, he brought expertise from a European perspective on sustainability transitions and environmental innovations (MSSI 2023). Under his leadership, one of the initial products of the 100RC, the *Resilient Melbourne Strategy (*2016*)* was Melbourne's first attempt[37] at metropolitan wide climate planning, as well as the first resilience strategy produced by any Australian city. Primary actions include a metropolitan wide bicycle network, and

[35] While large scale philanthropy is more common in the United States, the funding model is unusual in the Australian context. Further, the social values of outspoken private partnerships with government reflect international values that are more aligned with global trends in towards cross-sector private involvement in urban planning and governance (Croese et al. 2020; Leitner et al. 2018).

[36] The City of Melbourne provides the project team and additional investment (Resilient Melbounre, 2016).

[37] The Strategy was developed between July 2015 and May 2016.

an extension to urban forest strategy called *Living Melbourne (2019)*, aimed at creating comprehensive vegetation coverage across the wider city[38] (2019, p. 22).

Discussed earlier in this chapter, the requirement to work across council areas of Greater Melbourne contradicts the existing local governance structure (Davidson and Gleeson 2018, p. 233).[39] Greater Melbourne is divided into 32 distinct local councils, each with unique portfolio divisions representing specific community needs. The major driving external influence of *Resilient Melbourne* introduced a broader scale perspective on local governance which in turn raised questions about how to create a culture of collaboration in a fragmented governance structure. While many of the inner city CoM climate policies such as *Climate Change Adaptation refresh (2017), Green Our City Strategic Action Plan* (2017c) and the *Climate Change Mitigation Strategy to 2050: Melbourne together for 1.5 °C* (2018a) emphasis greater cohesion between policy objectives, these are particular to the relatively small inner city and not the more regional scale approach encouraged by 100RC.

The new policies of *Resilient Melbourne* and *Living Melbourne* were highly innovative in addressing these gaps, by incorporating methodologies for surveying, and combining diverse community perspectives in the policy making process.[40] For example, *Resilient Melbourne* was developed through a series of workshops across 12 months involving more than '1000 individuals, from 230 public and private organizations, Melbourne's 32 councils, and many state government departments'(Moloney and Doyon 2021, p. 5). Similarly, *Living Melbourne* utilised workshops[41] and meetings as active consultation where the 'varied stakeholders' priorities and their existing 'yet fragmented and often ad hoc' actions were identified and brought together (Coenen et al. 2020, p. 146).

These methods are considered highly experimental in the Australian context (Coenen et al. 2020, p. 146), not only for drawing together a greater diversity of voices but for positioning policy making as an open ended processes of ongoing development. For example, *Living Melbourne* is envisaged as a 'continuous process rather than an end in itself' (p. 147) where the multi-pronged policy aims to not only *provide guidance* but also to offer an ongoing 'self-organising' mechanism for drawing community leaders together by 'creating spaces for learning and reflection within the process' (Coenen et al. 2020, p. 146).

---

[38] Living Melbourne further identified a need for adding private land and developer-led urban greening to the original public space canopy targets.

[39] Davidson and Gleeson (2018) state; 'Resilient Melbourne represents an *urban governance experiment* in governing for resilience, providing an urgently needed contribution to our understanding of governance characteristics that influence action across complex urban areas.'

[40] collaboration is necessary to 'tackle both the 'acute shocks' and 'chronic stresses' that linked to natural and anthropogenic hazards as well as socio-economic and socio-technical challenges of the growing city'(Fastenrath et al. 2019, p. 693).

[41] As the *Living Melbourne Strategy (2019, p.9)* states 'only by working collaboratively between organisations, land tenures, and regions can we understand the needs, agree on priorities for protection and improvement, set targets, and track progress.'

The collaborative and process-driven models underpinning *Resilient*, and *Living Melbourne* therefore represent a radical shift in the norms of Australian policy making—far more representative of innovations in urban climate resilience and policy from Europe. As this chapter has briefly traced, policy on climate and urban heat across Melbourne has diversified considerably since the initial Climate adaptation plan in 2009, now including a breadth of strategy alongside guidelines and leadership positions.

4.4.4 From Models of Policy Making to Tools for Designing in a Hotter Dryer Climate

Though this evolution, climate guidance has shifted from the dominant approach of relying on a direct translation of scientific research into policy targets, standards, and regulations. Dunn et al. (2018, p. 144) describes this approach as a 'loading dock model', where the interactions between science and policy are 'linear, rational processes—a unidirectional model of knowledge transfer'. Here, policy and guidance are connected by a straight line between 'facts' and 'values' under the assumption that science will provide objective facts which will then inform policy makers values, aligned by the shared principle of reducing variables and assessing risk (Elliott 2018, p. 2). Further positioning this, Dunn et al. (2018, p. 148) state 'the scientific method and policy-process both stem from long traditions with deeply embedded normative thinking'.

That *loading-dock model* at work is evident in the earlier climate policy making processes which directly drew from the scientific research institutes at local, state, and national levels. For example, the initial Climate Change Adaptation Strategy (2009) utilised research on adaptation pathways emerging from the Federal Department of Climate change. Similarly, the early urban heat research Cool Roofs Guide (The University of Melbourne 2011) is a direct translation of building science research conducted at the University of Melbourne to a set of guidelines on the construction and colour of roofs for the City of Melbourne (City of Melbourne 2012c). This guide offers a clear example of the loading-dock model in action where a discreet scientific focus on one aspect of the built environment is transposed to a direct outcome, isolated from other influential research areas such as community vulnerability to heat (Hot Spots mapping, 2009) or urban greening.[42]

In contrast to single research streams, the widely acknowledged successful Urban Forest Strategy (2012) was underpinned by the cross-institutional support from national institutes ARCUE and CSIRO, combined with the University of Melbourne and Monash University. Maintaining these particular environmental science partnerships form one of the core actions of the Strategy (City of Melbourne

[42] Even while acknowledging concurrent research in Green Infrastructure, cool roofs are specifically positioned as an alternative to urban greening (The University of Melbourne 2011).

2012a, p. 48) thereby ensuring a direct link between science and policy. However, while multi-partnered, other equally important influences for considering an urban forest such as urban and landscape design are absent, again reflecting a clear prioritising of science in this form of policy making.

The limited intersections across research institutes, or multidisciplinary expertise highlights the underlying major challenges in interdisciplinary planning and design in the Australian setting. Successful interdisciplinary models require both appreciations, understanding and transference of methods from one discipline to another. Yet the prevalence of discrete climate research centres, each with their own emphasis i.e. urban ecology (ARCUE), climate adaptation (VCCCAR), climate science (CSIRO) (NCCARF) suggests a culture of scientific camps and focus within just one disciplinary setting. Each of which are premised on, and 'depend upon and endorse a certain depiction of objective, value free authority' (Brown and Havstad 2017, p. 72).

This separation of research therefore is challenging when departing from the loading dock model towards more collaborative and multidisciplinary models in policy making, like co-design,[43] where diverse voices and values are deliberately sought out.[44] For example, *Living Melbourne* utilised experimental co-design processes with the aim of creating more adaptive planning outcomes through alliances between councils, community groups and technical experts. The policy making process itself was *the method* for extracting stakeholders needs, as well as the organising device for synthesising contrasting ideas within the strategy document (Coenen et al. 2020, p. 146). In this model, policy makers are repositioned as facilitators working across varied stakeholders to align diverse sociological values in the institutional setting of local government. And while *resilience* offers a cohesive overarching view for this model of collaboration, the policy is an artefact which holds those values, rather than a method for taking those actions into communities. So, as well as negotiating across the complex and often competing urban issues and interests, there is the major challenge of establishing and implementing priorities back into specific places.[45] As Fainstein (2018) surmises through reviewing 100 Resilient Cities New York, overarching terms like resilience can 'obscure distributional impacts'.

The remaining question of implementation hangs over the status of the *Resilient* and *Living Melbourne* planning strategies. While the increasingly collaborative policy making efforts have positioned policymakers as facilitators for negotiating

[43] Numerous scholars from across the social sciences have pointed towards a necessary shift away from a singular prevailing system of linear logic as the core mode for exploring complex sustainability challenges (Hadorn et al. 2006; Haraway 1988; Max-Neef 2005; Sardar 2010; Schmidt and Neuburger 2017; Tobi and Kampen 2018).

[44] Co-design is premised on collating and synthesising perspectives without a hierarchical ordering (Blomkamp 2018).

[45] Moloney and Doyon (2021, p. 10) further suggest 'In the Australian context, without the strong support and leadership of state governments, the capacity to implement metropolitan wide initiatives, particularly infrastructure and land use changes will remain limited.

diverse priorities—a far cry from producing blanket rules and regulations—does the application of the policy objectives into action still hinge on these central roles? Even though the intention behind *Living Melbourne* was to create a self-organising network for future policy development, the process draws attention to the influence of individuals in the policy development.

In this case, the introduction of co-design methods as central to policy making aligns with the appointment of Professor Coenen to the Chair of Resilience. His disciplinary perspective and experience coming from Europe brought fresh sociological values on climate action into the Australian setting. In addition, his expertise in sustainability transitions incorporate interdisciplinary theory and methods on developing and managing power relationships in institutional networks(Coenen et al. 2012). Approaches which are clearly reflected in the radical change to the policy making process.

The major influence of the Chair of Resilience position suggests the critical impact of the skill of an individual as a significant driver of the success of policy. In interdisciplinary theory, these roles are sometimes called *integrators* or *boundary spanners*, as specific people who work across professional and community sectors to encourage collaboration and trust (Hamilton et al. 2021, p. 462). While research suggests that these roles are critical to successful negotiations in complex decision making (Posner and Cvitanovic 2019), it also highlights a major vulnerability in terms of sustaining momentum and change. What happens if these individual actors leave or change roles?

In this light, the methods for managing complex dynamics and relationships in producing *Resilient-* and *Living Melbourne* have been critiqued for failing to properly address the nuances of relationships within Melbourne's social and institutional setting. Particularly the power dynamics already at work in the urban environment. Here Moloney and Doyon (2021, p. 10) explain:

> While seeking to empower Mayors and Council leaders may be one approach to enabling resilience, it is unlikely to challenge entrenched power imbalances or address the root drivers of vulnerability.

Thus, beyond governance, other forms of institutionalised power structures in Melbourne's urban environment which also define how policy hits the ground. For example, two thirds of the City of Melbourne's physical space is controlled by private interests, which are frequently shaped by what Rogers and Gibson (2021, p. 526) describe as 'unsolicited urbanism'. This describes the neoliberal processes of decision-making informed by financial powers and vested interests at the exclusion of 'pre-existing statutory planning norms, local-scale municipal planners, and the general public.' As a set of sociological values, such financially led development is often unstated and indirectly addressed, processes which sit uncomfortably with the ambitions of open and inclusive collaborative design.

Although *Living Melbourne* specifically identifies the need to work across public and private domains, the policy making struggles to address these informal power dynamics. Thus, creating a further point of tension between policy agendas and the influence of roles and relationships in the built environment which leave higher-level policy ungrounded and unclear in how it manifests in real sites and communities. Further highlighting the absence of the voice of urban and landscape designers in Melbourne's major climate and urban heat initiatives. The design disciplines which enact policy decisions most visibly as constructed public spaces are often positioned solely at the very end of both the traditional processes of loading-docks, which rely on science and emerging collaborative planning processes like co-design which instead focuses on bringing community voices into planning. Here, returning to the challenges of implementation and the need to ground policy objectives in real spaces has seen the relatively recent concurrent branch of institutional guidance on urban heat.

4.5 An Emerging Toolkit

From the late-2010s, a series of *design support tools* have emerged as a further significant step in bridging between science, policy, and design as researchers and policy makers increasingly recognised the need for 'rigorous, evidence-based tool customised for *local conditions,* which could contribute towards meeting a range of its policy objectives, as well as environmentally sustainable design targets' (J. Bush et al. 2021, p. 22). Aimed at translating science and data specifically for use by built environment professionals, these tools offer significant links for bringing science related to Melbourne's local climate into the design processes, but without the previously linear unidirectional models of knowledge transfer.

As discussed earlier, the earliest guides for urban heat took the form of Green Infrastructure (GI) which was tightly focused on how to successfully grow plants on buildings, roofs, and streets. However, these tools were of limited value for designers addressing urban heat given the strategies only considered the addition of vegetation. Absent from these guides was any consideration or direction on how the siting, orientation, and massing of urban form, along with the use of materiality might influence heat effects. The demand for more design supportive tools emerged from the feedback from design professionals such as architects, landscape architects, urban ecologists and strategic planners who were engaged in conversations with policy makers and researchers about Melbourne's climate challenges. Managing heat was still perceived as a major technical hurdle—and those working at the practical end of implementation sought straight forward design tools which translated complex scientific language and principles into design applications with demonstrated effects(Ding et al. 2019; Hartigan et al. 2021; LCL 2017a, b; Parsaee et al. 2019).

Following these recommendations, three key support design tools have been developed since 2017. Two of these projects emerged nationally, the Low Carbon Living projects *Microclimate and Urban Heat Island Mitigation Decision-Support Tool (*2017c*)* and Australia's first *Guide to Urban Cooling Strategies (*2017b*)* while the *Green Factor Tool* (2020) was developed by the City of Melbourne. This final section will briefly discuss the success of these three tools in two ways: first in their value in providing guidance to designer's for mitigating urban heat in outdoor spaces in Melbourne, and second in their efficacy as multidisciplinary bridges between research, policy, and design outcomes.

4.5.1 Microclimate and Urban Heat Island Mitigation Decision-Support Tool (2017)

This decision support tool was developed by researchers from The University of New South Wales in Sydney, (UNSW) and Swinburne University in Melbourne, aligned under the Low Carbon Living Urban Microclimate project. The web-based tool takes an urban precinct (multiple urban block) scale approach to revealing the effects of urban heat by combining data from weather stations and local temperature sensors with planning data sets and models of urban form. With the addition of relevant data, the tool can be adapted for any Australian city. The visual interface can then be used to test temperature scenarios for specific blocks based on the cooling effects of building heights, cool roofs, hard materials, and optimised urban greenery (Ding 2018).

By allowing for simple simulations of heat effects over time, the model offers a data-driven view of urban heat build-up and retention across combined decisions about urban form and materials. This allows for planning and pre-design decisions on things like the size of urban blocks, the massing of buildings and the impact of vegetation, all considered against the potentials of future urban warming. The aim, described by the project developers Ding, Petersen, and Craft (2019, p. 3) is:

> to provide governments and built environment industries with a decision-support tool to inform urban policy, development assessment and planning practices related to potential building and urban interventions, used to cool streetscapes and cities, decrease energy consumption, protect the population's vulnerable health-wise, and improve conditions of comfort.

In Australia, this represents a major shift in how digital tools and environmental data provide guidance directly for design decisions about urban heat. Rather than using the data to evidence existing conditions and prove heat effects at whole city scales,[46] the tool uses data to expose local conditions through a series of future possibilities, allowing for planners and designers to simulate combinations of major heat influences. This encourages a far more complex and integrated design response than relying on single interventions like a green roof or a tree.

However, the *Microclimate and Urban Heat Island Mitigation Decision-Support Tool* maintains accuracy by restricting the available variables and simplifying systems. Because the model is intended to provide fast feedback using the most

---

[46] The project developersDing, Petersen, and Craft (2019, p. 6) further explain; 'to date much of this research has been conducted on large scale domains and is unsuitable to support practical development guidelines at a precinct and building level.'

well-established modes of managing urban temperatures the model can only show the potential for cooling through the pre-determined set of variables, i.e., modification of building height, the adoption of generic cool roofs and materials and optimised greening arrangements. So, while powerful for simulating heat effects, the mapping sits between analytical and very early-stage design processes. The coarseness of the interventions restricts its potential to planning scale decisions with little value for more detailed design decisions. In recognising a need to communicate even more detail about urban design interventions for managing heat, the same, scientific research used for this project was further translated into a complementary guideline document for urban cooling.

4.5.2 *The Guide to Urban Cooling Strategies (2017)*

These guidelines aim to synthesise the scientific urban microclimate research into an even more 'user-friendly' and practical document for built environment professionals (LCL 2017b, p. 4). Rather than the scenario based feedback of the interactive model, this document provides a list of evidenced street-scale interventions for creating cooling effects. These include shade and misting infrastructures, cool surfaces like permeable paving as well as applying vegetation through street trees and green roofs.

The effectiveness of each type of intervention is given a rating (1–3) that is specific to the climate of each major Australian city. For example, the table for Melbourne (Fig. 4.10) shows that green roofs, tree canopies and evaporative cooling

LOCAL CLIMATE				COOL PAVING			COOL ENVELOPE		GREEN ENVELOPE			EVAPORATIVE COOLING		
SUMMER: HOT CONTINUAL	WINTER: COOL	RECORD TEMPERATURE	AVERAGE RAINFALL (MM/Y)	HIGH ALBEDO PAVING	HIGH EMITTANCE PAVING	PERMEABLE PAVING	HIGH ALBEDO ENVELOPE TREATMENTS	HIGH EMITTANCE ENVELOPE TREATMENTS	GREEN ROOF	GREEN WALL	TREE CANOPY	SURFACE WATER AND EVAPORATIVE COOLING	MISTING FAN	SHADING STRUCTURES
		Min -2.8°C Max 46.4°C	650	1	3	3	R-3 W-1	RW-3	3	3	3	3	3	3

Effectiveness
High = 3 Low = 1
Medium = 2

W = Wall
R = Roof

Fig. 4.10 The effectiveness rating, between 1 and 3, of urban cooling interventions for Melbourne in the CRC Guide to Urban Cooling Strategies (2017b, p. 49). For example, high albedo paving has low effectiveness, whereas permeable paving has high effectiveness. (Illustration by Dr. Carlos Bartesaghi Koc)

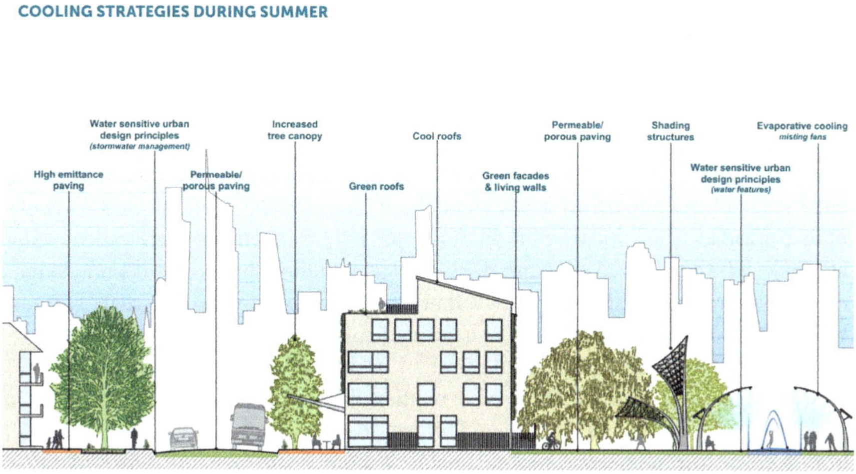

Fig. 4.11 The section for Melbourne, showing the different urban cooling tactics in a potential streetscape from the CRC Guide to Urban Cooling Strategies (2017b, p. 49). (Illustration by Dr. Carlos Bartesaghi Koc)

all have a high rate (3) of effectiveness whereas high albedo paving is low (1). The guide also includes sectional diagrams (Fig. 4.11) depicting how the various strategies might fit together into a generalised urban environment. For instance, how the shape and scale of interventions like trees, shading structures or evaporative cooling devices might work across streets, buildings, and outdoor spaces.

Like the decision-support tool, these guidelines represent an important shift in acknowledging the variable effectiveness of cooling that is unique to different geographic locations. While not able to offer the same data-driven feedback of the decision-support model, the document is a rigorous synthesis of recent scientific urban climate research. Although the *Guide to Urban Cooling Strategies* is technically detailed in terms of aligning the body of existing environmental science on urban cooling with potential interventions within the varied local climates of Australian cities, this has major epistemic consequences for design.

The guidelines maintain a deliberately tightly focus on mitigating heat—opposed to working with temperatures. By distilling heat to a single factor problem, further suggests heat can solved by set of pre-defined technologies. For example, airborne misting fans (evaporative spray systems) cool people down as fine water particles and air flow combine on human skin to produce a relative cooling effect. While effective, the approach encourages generalisable knowledge and creates 'rules of thumb' rather than methods for working with the behavioural traits of temperature. Or for conceiving of how the qualities of the atmosphere and climatic effects can be applied as innovative design working with atmospheres. Without methods for understanding local and site-specific variations and behaviours, or how to test these principles in specific sites, the generalised guidance has a problematic relationship with design. By specifying a set of technical solutions, the guidelines point to predetermined outcomes -a kit of part responses.

Developed from the same National CRC project, both the *Guideline*s and *Decision Support tool* share the ambition of identifying and solving the issue of heat in cities. As important pieces of institutional guidance, they have gone some way to bridging the gap between science and design, communicating environmental science about mitigating urban heat to local built environment professionals. However, both prioritise a technical solution for heat, underpinned by sociological values aligned with the rational 'loading dock model' which assumes that if knowledge is available, it will be utilised. Further, that scientific objectivity can create solutions which are the epistemic pathway to a defined solution.

This pattern is not unique to Australian policy setting. Across urban microclimate research and design, there is a methodological tendency towards reducing variables to generalisable outcomes. For designers, this sets up a barrier to application because the predetermined technical responses are general, whereas design application is specific to sites. Further, the simplification of the technical response can rarely be interpreted cleanly in a multi-factor design project. For example, in the case of Melbourne, 6 months of the year (or more) experiences cold weather. By focusing on solutions (epistemic) explicitly for heat, the guidelines and support tools deliberately avoid other climatic conditions. While the problems associated with heat and urban warming are undoubtedly significant for Melbourne, interventions that are exclusively developed for heat ignore the more complete lived experience of the seasonal cycles of the city.[47]

An emphasis on single technical solutions therefore creates obstructions for how designers conceptualise heat in design projects. While the Guidelines and Decision-support tools demonstrate how to manage heat, somewhat surprisingly they do not build an argument for why working with urban cooling effects is necessary for built projects. This is highlighted in the *Cooling Cities Strategies and Technologies to Mitigate Urban Heat discussion paper (*LCL 2017a, *p. 34),* which states:

> the issue of urban heat is not considered important for health and well-being of the community but rather a cost on development.

The final tool discussed here addresses this gap by focusing on the benefits and value of increasing environmental performance in built environment projects. While still working with data and environmental science, the *Green Factor Tool* also draws on process-driven collaborative models and the epistemic values of holism in addressing urban heat as part of the built environment's climatic performance.

---

[47] Or as Porter, Hurst, and Grandinetti (2020, p. 8) comment 'the tendency to reduce urban nature to a policy fix, stripped of wider political and cultural concerns.'

4.5.3 *The Green Factor Tool (2020)*

The Green Factor Tool is a City of Melbourne local council project[48] aimed at increasing urban greening—a direct connection to the earlier policy focus on Green Infrastructure. Available as a free web-based interface, the tool provides a scorecard (Fig. 4.12) with supporting analysis for development that calculates the inclusion of trees, groundcovers, shrubs, climbers, and green roof vegetation over seven environmental and social benefits. These include functional environmental performance for habitat, surface runoff, and temperature regulation but also social benefits of food supply, recreation, and aesthetics shown as parallel targets (Fig. 4.13). While calculated separately, these are positioned as concurrent goals. Shown below, the existing measurement is in green, which is then compared against a target performance-shown as the blue dashed circle.

As with the Decision Support tool and Guidelines discussed above, the *green factor tool* is aimed at built environment professionals including 'landscape architects, architects, environmentally sustainable design consultants' (Bush et al. 2021, p. 28) who work on, and influence public and private urban realms.

In a key difference to the Microclimate decision support tool which presented data-driven simulations of heat as open-ended scenarios, the *Green Factor Tool* suggests best practice performance targets.[49] By including targets and priorities, the

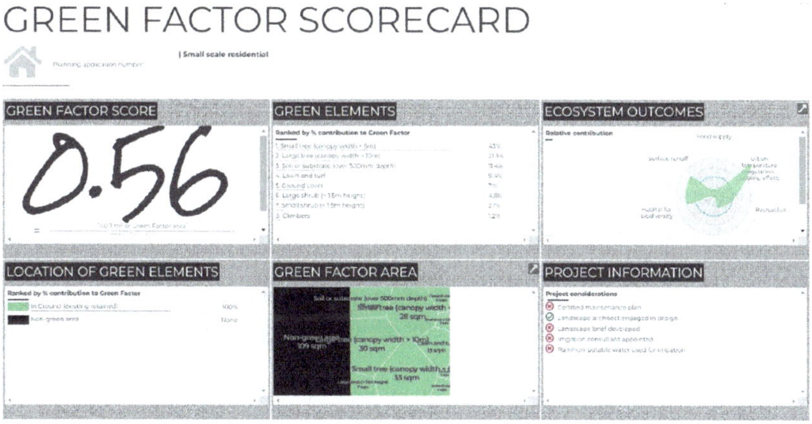

Fig. 4.12 Example Green Factor scorecard for a small residential typology, showing a score of 0.56 (City of Melbourne 2020)

[48] The project extending from and was supported by the University of Melbourne researchers who had contributed to the earlier Low Carbon Living tools. The project 'consultancy team appointed by City of Melbourne to develop the tool was led by HIP V. HYPE Sustainability, with tool and website design by Little Sketches, and research input from University of Melbourne researchers.' (J. Bush et al. 2021, p. 22)

[49] 0.55 was set as the target Green Factor Score for residential and commercial developments (corresponding to a horizontal green cover of 40% site coverage), and 0.25 for industrial developments (Bush et al. 2021, p. 26).

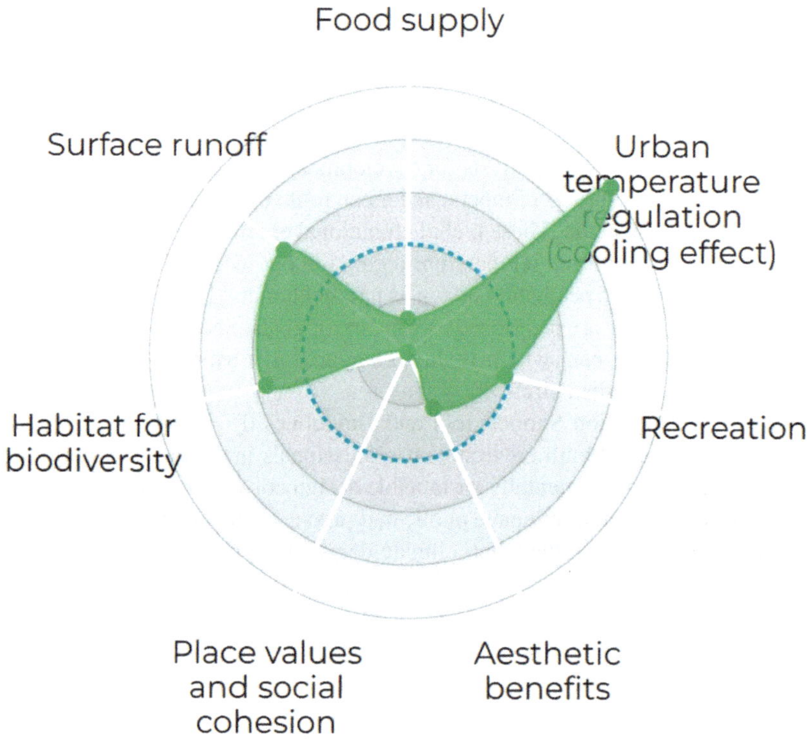

Fig. 4.13 An example showing the comparative measures of 'ecosystem outcomes'. The measurement shown in green for each factor can also be compared to the target of 0.55 which is represented by the blue dotted circle (City of Melbourne 2020)

Green Factor Tool offers a further notable shift in design support tools away from proving neutral information, towards engaging with the social values which underpin practical decisions about climate performance. Rather than assuming value-free provision of information, the tool directly argues for greater ecosystem performance through the inclusion of greening.

In doing so, the tool takes on the issue of who determines values and priorities as a multidisciplinary question. As discussed, the conventional loading-dock model of science advising assumes that science provides objective information and that policy makers will determine the values in how this is applied. In contrast, The *Green Factor Tool* employs a transdisciplinary framework (Bush et al. 2021, p. 27) bringing together disciplinary influences from planning, design and science to negotiating and determining the priories of the tool. The project developers describe a six-stage process of research, consultation, peer review and piloting with disciplines of strategic and statutory planning, urban design, open space planning, urban landscapes management, landscape architecture and urban ecology. Across the stages of consultation, academics, practice professionals and council staff were invited to workshops,[50] discussions and testing to determine the criteria, prioritizing of functions, rating system and setting the target score(Bush et al. 2021, p. 26).

Like the co-design process discussed about in the creation of *Living Melbourne* strategy the process of creating the tool is key to the method of facilitating between disciplinary perspectives, which is managed and controlled by the researchers on the project, and again, the artifact of the process holds then the values. Even though these have been determined through the collaborative process, these are now established and held within the tool through the determined criteria, benchmarks, and targets. Strategically, these positions where the tool might be used in the planning stages of a development process where project ambitions are being set. While the feedback on specific sites, includes the very fine detail of the size and type of vegetation, this operates as a guideline rather than spatial design tool. As Melbourne based landscape architect Anna Feldman (2022) summarises:

> the scorecard is useful for leveraging an argument as a tool for negotiation to make sure landscape is given due consideration but doesn't carry through to design.

Here, coming back to the earlier questions of implementation and how to apply this kind of knowledge into design projects returns to the core challenge of how the diverse knowledges required to address climate issues work together in design.

The Green Factor Tool is a powerful instrument in supporting the priorities which are embedded into it but as a support tool, does not extend into the design process. Equally, it does not address the relationships of the built environment setting and institutions which carry concurrent and even competing values and priorities. As this chapter has outlined, Australian urban governance is fractured and complex, which Judy Bush (2020, p. 42) describes as:

> an absence of shared or overarching narrative; the 'splintered narratives' reflect inward facing competition between advocates of green spaces' different functions and ecological composition.

Although the tools described here each fill some of the foremost gaps in equipping planners and designers with guidance, scientific knowledge, and site feedback, these exist within complex, and often disjoined institutional frameworks.

At the time of writing there is discussion about making the *Green Factor Tool* mandatory for new developments within the City of Melbourne. The intention of the researchers and tool developers, Bush et al. (2021, p. 26) is to 'monitor the tool's use by designers and developers during the voluntary phase and pursue changes to the Melbourne Planning Scheme that would integrate an assessment metric for greening targets'.[51] The setting of base level targets for urban greening would bring

[50] Workshop one was with one landscape architecture researcher; workshop two was with three urban ecology researchers; workshop three was with more than 10 green infrastructure and urban ecology researchers; and workshop four was with four Council staff from the landscape and planning teams (Bush et al. 2021, p. 24).

[51] At the time of writing the website states 'The tool is currently voluntary, but we are asking all developers who submit a planning application for new buildings (except single dwellings) to submit a green factor scorecard, aiming to achieve a target score of 0.55 or 0.25 for industrial. This

Melbourne's institutional response closer to successful examples like Singaporean Green Plot Ratios, which determine green vegetation to hard-surface ratios in new developments. However, the success of Singapore's urban greening program lies in the integration of planning controls alongside strong institutional guidance and direct incentives (Shi et al. 2017, p. 2). For example, the Singaporean government offers further incentives to developers including allowing building height limits to be raised when a development incorporates more rooftop greening.

While the three Australian urban heat design support tools discussed here are freely and publicly available, they still exist in isolation without any further incentive or regulation for applying them to new developments or design projects. This is pronounced in the private sector, which controls two third of Melbourne's urban space, and is lacking substantial motivation to do more than the minimum in responding to urban heat and climate effects, particularly when weighed against competing demands of budgets and timelines(Dunn et al. 2018; Hamilton et al. 2021; LCL 2017a, p. 34).[52]

For the built environment, the gap between the ambitions of institutional guidance and how work is implemented represents the foremost hurdle in driving policy-led climate initiatives through design outcomes. The lineage of policy making, models of collaboration, relationships and power structures create a complex push and pull of priorities and agendas where the design professions are just one aspect of the more extensive network of decision-makers in the urban climate.

4.6 Conclusion

This chapter has traced the challenges of Australian urban climate governance, beginning with the weather patterns unique to the City of Melbourne and the challenges of working with an urban form influenced by a colonial heritage that does not account for the local climate. Further, the colonial cultural background which has established ongoing sociological views of the environment, such as the concept of enduring heat as an Australian characteristic. These factors feed directly into politics—particularly in the Australian democratic system where the three-tier system of separately elected Federal, State and Local governments represents a cross-section of social worldviews. This analysis has shown that climate governance in this system is fragmented across concurrent portfolios and agencies, which creates significant underlying challenges in bringing together research, urban planning, and design. The analysis also shows, at the smaller scale of local government, extreme weather events like heatwaves and drought have been transformative in moving

will help developers to get familiar with the tool and understand future requirements (City of Melbourne 2020).

[52] For example Hamilton et al. (2021, p. 468) observes 'At present there are few incentives to broker the existing collaboration gaps in the complex and competitive nature of the urban environment.'

local policy response from tentative risk assessment towards more detailed responses to Melbourne's heat. With the rolling changes to external funding and data-enabled technology, alongside the socio-political model of the Resilient Cities project, local climate planning has turned towards increasingly holistic and collaborative models of policymaking.

Through this story, the role of policymakers has evolved towards having greater agency in directing Melbourne's urban climate response. But this is not a single disciplinary story—Melbourne's policymaking draws on knowledge from environmental scientists, planners, economists, engineers, and designers. Despite significant developments in strategic planning through initiatives like *Resilient-* and *Living Melbourne*, the implementation gaps between environmental science and designing built projects have remained persistent challenges to actioning knowledge. Here, the most recent institutional guidance has taken on the form of design support tools, offering data-driven mapping, targets, and translation of science, especially for planners and designers to bring climate and heat mitigation to the fore of design projects.

Looking across this genealogy of climate policy, it's evident that the collaboration models behind Melbourne's policy response still favour environmental science perspectives, which have profound epistemic consequences on how policy initiatives carry through to design. In Melbourne, environmental data and urban heat simulations have informed canopy targets and design guidelines which continue to privilege and value technical responses, further enforcing an epistemic pathway of generalised approaches. While it is assumed that spatial design will easily follow this kind of guidance, these are still far from offering design performance principles. The influence of policy, and the institutions and relationships which produce it are major factors in how the warming climate is perceived and responded to by designers operating at the coal face of a project. Design, as the detailed making of space, remains at arms-length from the planning and assessment of heat. So, although the more recent design tools are significant steps towards communicating and negotiating with developers and designers, they are still positioned as voluntary guidance that is separate from the core business of design in built environment projects.

References

ABC (2022) Australia's political system. Retrieved from https://www.abc.net.au/ra/federasi/tema1/aus_pol_chart_e.pdf

ABS (2021) Statistics people population Victoria https://www.abs.gov.au/statistics/people/population/regional-population/2020-21#victoria

AECOM, N (2013) Supporting evidence-based adaptation decision-making in Victoria: a synthesis of climate change adaptation research. https://nccarf.edu.au/wp-content/uploads/2019/03/AECOM_2013_Synthesis_report_for_VIC.pdf

Banwell C, Dixon J, Bambrick H, Edwards F, Kjellström T (2012) Socio-cultural reflections on heat in Australia with implications for health and climate change adaptation. Glob Health Action 5(1):19277

Blomkamp E (2018) The promise of co-design for public policy 1. In: Routledge handbook of policy design. Routledge, pp 59–73

BOM (2009) Special climate statement: the exceptional January–February 2009 heatwave in south-eastern Australia. Bureau of Meteorology: Australian Government

BOM (2017) Special climate statement 63 – a prolonged warm spell in Tasmania and Victoria. Bureau of Meteorology: Commonwealth of Australia

BOM (2023) Bureau of meteorology map of climate zones of Australia. Retrieved from http://www.bom.gov.au/climate/maps/averages/climate-classification/?maptype=kpn

Boschat G, Pezza A, Simmonds I, Perkins S, Cowan T, Purich A (2015) Large scale and sub-regional connections in the lead up to summer heat wave and extreme rainfall events in eastern Australia. Clim Dyn 44(7-8):1823–1840

Brennan MW, Cleary W, Cleary H (2011) Local adaptation pathways program forum report. Retrieved from https://www.pc.gov.au/inquiries/completed/climate-change-adaptation/submissions/sub038-attachment1.pdf

Brown MJ, Havstad JC (2017) The disconnect problem, scientific authority, and climate policy. Perspect Sci 25(1):67–94

Bush J (2020) The role of local government greening policies in the transition towards nature-based cities. Environ Innov Soc Trans 35:35–44

Bush J, Ashley G, Foster B, Hall G (2021) Integrating green infrastructure into urban planning: developing Melbourne's green factor tool. Urban Plan 6(1):20–31

City of Melbourne (2010) Heatwave response plan. City of Melbourne

City of Melbourne (2012a) Urban Forest Strategy: Making a Great City Greener 2012–2032. Retrieved from https://www.melbourne.vic.gov.au/SiteCollectionDocuments/urban-forest-strategy.pdf

City of Melbourne (2012b) Open space strategy planning for future growth. Retrieved from https://www.melbourne.vic.gov.au/openspace-strategy

City of Melbourne (2012c) Cool roofs: a city of Melbourne Guide. Retrieved from https://www.melbourne.vic.gov.au/SiteCollectionDocuments/cool-roofs-guide.pdf

City of Melbourne (2017a) Climate change adaptation strategy refresh 2017. http://www.melbourne.vic.gov.au/about-council/vision-goals/eco-city/pages/climate-change-adaptation-strategy.aspx. City of Melbourne

City of Melbourne (2017b) Plan Melbourne. The State of Victoria Department of Environment, Land, Water and Planning.

City of Melbourne (2017c) Green our city strategic action plan 2017–2021

City of Melbourne (2018a) Heatwaves. Retrieved from http://www.melbourne.vic.gov.au/community/safety-emergency/emergency-management/Pages/heatwaves.aspx

City of Melbourne (2018b) Respite for homeless during extreme hot weather. Retrieved from http://www.melbourne.vic.gov.au/news-and-media/Pages/respite-for-homeless-during-extreme-hot-weather.aspx

City of Melbourne (2020) Green factor tool Retrieved from https://www.greenfactor.com.au/

City of Melbourne (2022a) Building Melbourne. Retrieved from https://www.melbourne.vic.gov.au/about-melbourne/melbourne-heritage/history/pages/building-melbourne.aspx

City of Melbourne (2022b) Parks and gardens history. https://www.melbourne.vic.gov.au/about-melbourne/melbourne-heritage/history/Pages/parks-gardens-history.aspx

City of Melbourne (Cartographer) (2012) Thermal map Melbourne city. Retrieved from https://data.melbourne.vic.gov.au/explore/dataset/thermal-image-2012/information/

City of Melbourne (Cartographer) (2015) Cool places – City of Melbourne. https://www.melbourne.vic.gov.au/about-melbourne/melbourne-profile/pages/city-maps.aspx

Coenen L, Benneworth P, Truffer B (2012) Toward a spatial perspective on sustainability transitions. Res Policy 41(6):968–979

Coenen L, Davidson K, Frantzeskaki N, Grenfell M, Håkansson I, Hartigan M (2020) Metropolitan governance in action? Learning from metropolitan Melbourne's urban forest strategy. Aust Plan 56(2):144–148
Cornes IC, Cook B (2018) Localising climate change: heatwave responses in urban households. Disaster Prev Manag 27(2):159–174
Coutts AM, Beringer J, Tapper NJ (2007) Impact of increasing urban density on local climate: spatial and temporal variations in the surface energy balance in Melbourne, Australia. J Appl Meteorol Climatol 46(4):477–493
Coutts AM, Tapper NJ, Beringer J, Loughnan M, Demuzere M (2013) Watering our cities the capacity for water sensitive urban design to support urban cooling and improve human thermal comfort in the Australian context. Prog Phys Geogr 37(1):2–28
Croese S, Green C, Morgan G (2020) Localizing the sustainable development goals through the lens of urban resilience: lessons and learnings from 100 resilient cities and Cape Town. Sustainability 12(2):550
Crowley K (2021) Fighting the future: the politics of climate policy failure in Australia (2015–2020). Wiley Interdiscip Rev Clim Change 12(5):e725
CSIRO (2007) Climate change in Australia technical report. Retrieved from https://climatechangeinaustralia.gov.au/en/communication-resources/reports/
CSIRO, B. o. M (2022) State of the climate 2022. http://www.bom.gov.au/state-of-the-climate/2022/documents/2022-state-of-the-climate-web.pdf. Bureau of Meteorology
D.o.H.S. Department of Health Services (2009) 2009 heatwave in Victoria: an assessment of health impacts. Victorian Government, Melbourne
Davidson K, Gleeson B (2018) New socio-ecological Imperatives for cities: possibilities and dilemmas for Australian metropolitan governance. Urban Policy Res 36(2):230–241
DCC (2009) City of Melbourne climate change adaptation strategy. Retrieved from http://www.melbourne.vic.gov.au/AboutCouncil/PlansandPublications/strategies/Documents/climate_change_adaptation_strategy.PDF
DCCEEW (2022) Climate adaptation in Australia. https://www.dcceew.gov.au/climate-change/policy/adaptation
DEECA (2023) More trees for a cooler greener west. Retrieved from https://www.environment.vic.gov.au/more-trees-for-a-cooler-greener-west/program/about-the-program
DEPI, Department of Environment and Primary Industries (2014) Growing green guide: a guide to green roofs, Walls & Facades in Melbourne and Victoria, Australia
Department of Environment, L., Water and Planning, DELWP (2016) Managing extreme water shortage in Victoria: lessons from the MIllenium Drought. https://www.water.vic.gov.au/__data/assets/pdf_file/0017/512720/DELWP-MillenniumDrought-web-SB.pdf.pdf. The State of Victoria
Department of Health, Victoria (2014) The health impacts of the 2014 heatwave. State Government of Victoria
Department of Health, Victoria (2015) Heat health plan for Victoria: protecting health and reducing harm from extreme heat and heatwave. State of Victoria
Ding LEA (2018) Microclimate and Urban Heat Island Mitigation Decision-Support Tool (UHI-DS). Retrieved from http://uhimitigationindex.be.unsw.edu.au/uhitool/index.html
Ding L, Petersen H, Craft W (2019) Microclimate and urban heat island mitigation decision-support tool (Project short report). Retrieved from http://www.lowcarbonlivingcrc.com.au/sites/all/files/publications_file_attachments/rp2023_project_short_report.pdf
Dow A (2015) Plans to use MCG as shelter for Melbourne's heatwave refugees. The Age. Retrieved from https://www.theage.com.au/national/victoria/plans-to-use-mcg-as-shelter-for-melbournes-heatwave-refugees-20150213-13em74.html
DTP (2023a) Cooling and greening Melbourne. Retrieved from https://www.planning.vic.gov.au/policy-and-strategy/planning-for-melbourne/plan-melbourne/cooling-greening-melbourne
DTP (2023b) Department Update. Retrieved from https://dtp.vic.gov.au/about/transport news/news-archive/department-update

Dubash NK (2021) Varieties of climate governance: the emergence and functioning of climate institutions. Environ Politics 30(sup1):1–25

Dunn G, Bos J, Brown R (2018) Mediating the science-policy interface: insights from the urban water sector in Melbourne, Australia. Environ Sci Policy 82:143–150

Elliott KC (2018) Roles for socially engaged philosophy of science in environmental policy. In: The Palgrave handbook of philosophy and public policy. Springer, pp 767–778

Fainstein SS (2018) Resilience and justice: planning for New York City. In: The resilience machine. Routledge, pp 159–176

Fastenrath S, Coenen L, Davidson K (2019) Urban resilience in action: the resilient Melbourne strategy as transformative urban innovation policy? Sustainability 11(3):693

Feldman A (2022) Working with landscape assessment tools in design/Interviewer: W. Walls. unpublished

Ferng J, Chang J-H, L'Heureux E, Ryan DJ (2020) Climatic design and its others: "southern" perspectives in the age of the anthropocene. J Archit Educ 74(2):250–262

Forino G, von Meding J, Brewer G, Gajendran T (2014) Disaster risk reduction and climate change adaptation policy in Australia. Procedia Econ Financ 18:473–482

Foster+Partners (2007) Masdar Development. Retrieved from http://www.fosterandpartners.com/projects/masdar-development/

Google (2023) Map data. Australia. Retrieved from https://www.google.com/maps/

Hadorn GH, Bradley D, Pohl C, Rist S, Wiesmann U (2006) Implications of transdisciplinarity for sustainability research. Ecol Econ 60(1):119–128

Haining V (2005) Tree management/replacement program, ompact and implications. Retrieved from http://www.melbourne.vic.gov.au/opm/bc/CTEE/meetings/PaE_53_20050531.pdf

Hamilton M, Fischer AP, Jasny L (2021) Bridging collaboration gaps in fragmented environmental governance systems. Environ Sci Policy 124:461–470

Hanna E (2022) Climate change—reducing Australia's emissions. https://www.aph.gov.au/About_Parliament/Parliamentary_Departments/Parliamentary_Library/pubs/BriefingBook45p/EmissionsReduction

Haraway D (1988) Situated knowledges: the science question in feminism and the privilege of partial perspective. Fem Stud 14(3):575–599

Hartigan M, Fitzsimons J, Grenfell M, Kent T (2021) Developing a metropolitan-wide urban forest strategy for a large, expanding and densifying capital city: lessons from Melbourne, Australia. Land 10(8):809

Heberger M (2012) Australia's millennium drought: impacts and responses. In: The world's water. Springer, pp 97–125

Hoddle R (Cartographer) (1837) Robert Hoddle's survey of the town of Melbourne in 1837

Imran H, Kala J, Ng A, Muthukumaran S (2017) An evaluation of the performance of a WRF multi-physics ensemble for heatwave events over the city of Melbourne in southeast Australia. Clim Dyn 50:2553–2586. https://doi.org/10.1007/s00382-017-3758-y

Ireland T (2003) "The absence of ghosts": landscape and identity in the archaeology of Australia's settler culture. Hist Archaeol 37:56–72

Jacobs SJ, Pezza AB, Barras V, Bye J (2014) A new 'bio-comfort'perspective for Melbourne based on heat stress, air pollution and pollen. Int J Biometeorol 58(2):263–275

Kent T (2016) Resilient Melbourne strategy: report to the future Melbourne (Environment) Committee. Retrieved from https://www.melbourne.vic.gov.au/about-council/committees-meetings/meeting-archive/MeetingAgendaItemAttachments/729/13230/may16-fmc2-agenda-6-9.pdf

Lau A (2012) Masdar City: a model of urban environmental sustainability. Stanford Social Sciences:77–82

LCL (2017a) Cooling Cities: Strategies and technologies to mitigate urban heat discussion paper. Cooperative Research Centers: Low Carbon Living. Retrieved from http://www.lowcarbonlivingcrc.com.au/sites/all/files/event_file_attachments/discussion_paper_cooling_cities_final.pdf

LCL (2017b) Guide to Urban Cooling Strategies. Cooperative Research Centers: Low Carbon Living: CRC Retrieved from http://www.lowcarbonlivingcrc.com.au/sites/all/files/event_file_attachments/crc_lcl_urban_cooling_guide_2017_web.pdf

LCL (2017c) Microclimate and urban heat island mitigation decision-support tool. CRC Low Carbon Living

Leitner H, Sheppard E, Webber S, Colven E (2018) Globalizing urban resilience. Urban Geogr 39(8):1276–1284

Lewis M (1995) Melbourne: the city's history and development. City of Melbourne

Loughnan, M., Nicholls, N., & Tapper, N. (2009). Hot spots project: a spatial vulnerability analysis of urban populations to extreme heat events.

Loughnan ME, Nicholls N, Tapper NJ (2010a) The effects of summer temperature, age and socio-economic circumstance on Acute Myocardial Infarction admissions in Melbourne, Australia. Int J Health Geogr 9(1):41

Loughnan ME, Nicholls N, Tapper NJ (2010b) When the heat is on: Threshold temperatures for AMI admissions to hospital in Melbourne Australia. Appl Geogr 30(1):63–69

MacNeil R (2021) Swimming against the current: Australian climate institutions and the politics of polarisation. Environ Polit 30(sup1):162–183

Max-Neef M (2005) Foundations of transdisciplinarity. Ecol Econ 53(1):5–16

McBride JL, Mills G, Wain AG (2009) The meteorology of Australian heatwaves. Paper presented at the modelling and understanding high impact weather: third CAWCR modelling workshop

Melbourne R (2016) Resilient Melbourne strategy. Retrieved from chrome-extension://efaidnbmnnnibpcajpcglclefindmkaj/. https://www.melbourne.vic.gov.au/sitecollectiondocuments/resilient-melbourne-strategy.pdf

Melbourne R (2019). Living Melbourne: our metropolitan urban forest. Resilient Melbourne. Retrieved from https://resilientmelbourne.com.au/living-melbourne/

Moloney S, Doyon A (2021) The Resilient Melbourne experiment: analyzing the conditions for transformative urban resilience implementation. Cities 110:103017

Moya R, Prohasky D, Watkins S, Ding Y, Burry J, Burry M (2013) Aerodynamic strategy applied in an urban shelter design. Paper presented at the eCAADe 32 towards smarter cities

MSSI (2023) Professor Lars Coenen. Retrieved from https://sustainable.unimelb.edu.au/about/honoraries-and-associates/honoraries/professor-lars-coenen

NCCARF (2013) Managing heatwave impacts under climate change: policy guidance brief 9 [Press release]

Nicholls N et al (2007) A simple heat alert system for Melbourne, Australia. Int J Biometeorol 52(5):375–384

Parsaee M, Joybari MM, Mirzaei PA, Haghighat F (2019) Urban heat island, urban climate maps and urban development policies and action plans. Environ Technol Innov 14:100341

Perkins SE, Argüeso D, White CJ (2015) Relationships between climate variability, soil moisture, and Australian heatwaves. J Geophys Res-Atmos 120(16):8144–8164

Pfeiffer S (2014) City and wind climate as an architectural instrument: DOM

Pittock B (2005) Climate change turning up the heat. CSIRO Publishing, Collingwood

Porter L, Hurst J, Grandinetti T (2020) The politics of greening unceded lands in the settler city. Aust Geogr 51(2):221–238

Posner SM, Cvitanovic C (2019) Evaluating the impacts of boundary-spanning activities at the interface of environmental science and policy: a review of progress and future research needs. Environ Sci Policy 92:141–151

Rockefeller (2013) 100 Resilient Cities. Retrieved from https://www.rockefellerfoundation.org/100-resilient-cities/

Rogers D, Gibson C (2021) Unsolicited urbanism: development monopolies, regulatory-technical fixes and planning-as-deal-making. Environ Plan A Econ Space 53(3):525–547

Saman W, Pullen S, Boland J (2014) How to cope with heat waves in the home. In: Palutikof P (ed) Applied studies in climate adaptation. Wiley, p 354

Sardar Z (2010) The Namesake: futures; futures studies; futurology; futuristic; foresight—what's in a name? Futures 42(3):177–184

Schmidt L, Neuburger M (2017) Trapped between privileges and precariousness: tracing transdisciplinary research in a postcolonial setting. Futures 93:54–67

Shaw K (2013) High rise Melbourne. Planning News 39(9):7

Shi Z, Fonseca JA, Schlueter A (2017) Building regulations and urban policies as incentives for application of district cooling systems in Singapore. Paper presented at the World Sustainable Built Environment Conference 2017

State Library of Victoria (2023) Melbourne city grid. https://ergo.slv.vic.gov.au/explore-history/colonial-melbourne/everyday-life/melbourne-city-grid. State Library of Victoria

Tapp R (1985) Patterns of weak, near-surface winds at Melbourne, Australia. Bound-Layer Meteorol 33(4):397–414

The University of Melbourne (2011) Cool roofs: City of Melbourne research report

Thrive (2017) Living well: apartments, comfort and resilience in climate change. Retrieved from Thrive Research: Green Cities Innovation Fund. The University of Melbourne

Tobi H, Kampen JK (2018) Research design: the methodology for interdisciplinary research framework. Qual Quan 52(3):1209–1225

Tranter B, Donoghue J (2007) Colonial and post-colonial aspects of Australian identity 1. Br J Sociol 58(2):165–183

Trounson A (2017) Why apartments are failing the heatwave test. Pursuit. Retrieved from https://pursuit.unimelb.edu.au/articles/why-apartments-are-failing-the-heat-stress-test

Trundle A, Bosomworth K, McEvoy D (2013) Combating the urban heat island effect: strategies for enhancing the cooling capacity of green infrastructure in Melbourne, Australia. Paving the Way Forward in Urbanization 9(July):33

Turton S (2019) The stubborn high-pressure system behind Australia's record heatwaves. January 25, 2019. The Conversation Retrieved from https://theconversation.com/the-stubborn-high-pressure-system-behind-australias-record-heatwaves-110442

UoM (2008) Melbourne Grid Plan. School of Historical & Philosophical Studies. Retrieved from https://www.emelbourne.net.au/biogs/EM00677b.htm

USDA Forest Service (2006) i-Tree Eco. Retrieved from https://www.itreetools.org/

VAGO, Victorian Auditor General Office (2014). Heatwave management: reducing the risk to public health. Retrieved from https://www.audit.vic.gov.au/sites/default/files/2017-07/20141014-Heatwave-Management.pdf

VEC (2022) Victoria electoral commission subdivision review. Retrieved from https://www.vec.vic.gov.au/electoral-boundaries/what-are-electoral-boundaries

Walls W, Parker N, Walliss J (2015) Designing with thermal comfort indices in outdoor sites. Paper presented at the living and learning: research for a better built environment: 49th international conference of the Architectural Science Association

Chapter 5
Maguey Landscape: Regeneration of the Desert

Rob Roggema, Esteban Sheridan, Miguel Ortiz, and Rodrigo Junco

Abstract This chapter explores the untapped potentials of cultivating the maguey plant, a species of agave widely found across Mexico and other regions. Historically, the plant's remarkable qualities have been overlooked or forgotten. Its most significant contribution lies in its ability to capture CO_2. The maguey plant is one of the most efficient processors of carbon capture in the world and can capture up to 65 tons CO_2 per hectare.

Beyond carbon capture, maguey offers several other benefits, including medicinal properties, the health benefits of *aguamiel*, soil improvement and ecological restoration of arid landscapes. However, unlocking this potential requires a large-scale, landscape planning approach to fully harness these benefits and contribute to reversing climate change.

The economic opportunities associated with maguey cultivation are substantial as well, as it offers local farmers the opportunity to sell products from the plant and capitalize on the carbon capture through carbon certification. By forming cooperative organizations, farmers can pool their knowledge and resources, amplifying their collective impact and ensuring the success of maguey cultivation on a broader scale.

Keywords Maguey · Arid landscape · Carbon capture · Soil improvement · Mexico · Organic agriculture

R. Roggema (✉) · M. Ortiz · R. Junco
Escuela de Arquitectura, Artes y Diseño, Tecnológico de Monterrey, Monterrey, Nuevo León, Mexico
e-mail: rob@cittaideale.eu

E. Sheridan
Saltillo, Mexico

R. Roggema (ed.), *Dry Urbanism*, Contemporary Urban Design Thinking, https://doi.org/10.1007/978-3-031-77347-1_5

5.1 Introduction

Global climate is changing at an unprecedented rate, which is very likely caused by a fossil-fuel supercharged economy, in combination with industrial agriculture and farms as factories. This leads to destructive land-use and consumption patterns, that have skyrocketed dangerous CO_2 in the atmosphere (Cummins 2020). In recent years the WMO has announced abnormal, record-breaking temperatures in large parts of the world during 2023 (WMO 2024), and continuing in 2024 (Climate Copernicus 2024). The way food is grown, the types of farming and current land use practice have decarbonized and reduced biological life of soils, forests and ecosystems, and their capacity to sequester carbon. More than 45,300 species are threatened worldwide with extinction, which is 28% of all assessed species (IUCN 2024). Oxidation in degraded and desertificated landscapes released lots of greenhouse gases and erased much of the above ground biomass. The natural soils, plants and forests can no longer capture and sequester enough CO_2 to maintain a balanced system. It is the main cause of global warming. More than 40% of the global land surface is currently arid or semi-arid. 15% of this is found in Latin America, mainly in Mexico (United Nations Convention to Combat Desertification XE "Desertification" 2019). The increased drought, erratic rainfall and degraded soils in these areas pose an enormous challenge on the users of the land (mainly farmers). There is a real risk these areas will degrade even further, ending up as a desert in which no cultivation is possible any longer (Cummings and Leu 2021). Already in 2003, in Mexico almost 45% of the soils are susceptible to some form of land degradation (Semarnat y CP 2003). It can be estimated this number has only increased in the last 20 years.

To prevent catastrophic climate change it is necessary to stop this degradation and restore soils and green and shade arid landscapes. This will have to lead to rapid increase of stored carbon, in the soil as well as above ground. As most of the degraded drylands are found in underdeveloped regions, where the most vulnerable populations are living, it is not merely a climate change but also a social and equity question that requires fast and fundamental solutions. Most farms and pastures do not have irrigation wells or sufficient access to surface water. Cultivation of crops is deteriorating and often trees and perennials are removed or burnt, and pastures are typically overgrazed. This vicious circle of degraded landscapes, limited resources, further degradation, poverty, no access to water, further degradation feels like a race to the bottom, until no human or natural life is possible anymore. However, the tide can be turned if local farmers and users of the land stop overgrazing and burning practice, but reforest, revegetate rehydrate and re-carbonate the soils by implementing traditional hydrological principles and organic agroforestry (ECPA 2022). One way of applying these principles is the use of agave/maguey cultivation and planting trees that fixate nitrogen.

5.2 The Maguey Plant—Plant of the Gods

There are approximately 200 species of agave worldwide, with 75% of them found in Mexico. About half of these species are endemic to the country. According to chronicler José de Acosta, the maguey plant is 'a tree of miracles' (De Acosta 1590), which gives water, wine, oil and vinegar, honey, *arrope (grape syrup) e hilo (thread), y aguja (needle)* and countless other resources. It is a generous plant which not only provides food but also fiber, textile culture, celebrations, and art. Moreover, the maguey is an iconic plant that shapes national identity, as it intertwines the environment, culture, and economy (Turok 2010). Alexander von Humboldt recognized it as the most useful plant that nature gave to the people in central America (Von Humboldt 1827).

The different varieties of maguey (Fig. 5.1), growing together with shrubby trees such as mesquite (Fig. 5.2), huizache (Fig. 5.3), desert ironwood, wattle and a range of acacias, all of which fix nitrogen deep in the soil through their extensive root systems, are among the most common yet overlooked plants in the world. "Agave (and maguey) is to the drier parts of the world what bamboo is to its wetter zones.

Fig. 5.1 Agave Salmiana (maguey)

Fig. 5.2 Mezquite

Capturing atmospheric CO_2 in vegetation is severely limited by the availability of land and water. The best choice would be species that can utilize lands unfit for food production and yet make the dynamics of carbon sequestration faster. As much as 40% land on earth is arid and semi-arid, largely in the tropics but also in the cool temperate zones up north. And on almost half of these lands, with a minimum annual rainfall of about 250 mm and soils that are slightly refractory, the very valuable species of agave/maguey grows reasonably well." (Kant 2010).

5.3 Maguey Agroforestry Systems

Densely, intercropped maguey and the accompanying trees have the capacity to capture large amounts of CO_2, while growing biomass under and above the earth's surface. It is an ideal crop for arid and semi-arid landscapes because the plants and trees require little or no irrigation or agrochemicals, and it thrives in poor soils (Cummins 2020).

Fig. 5.3 Huizache

On average, 2000 maguey plants and 400 mesquite trees can be planted per hectare. Each hectare of this system can capture the dry weight equivalent of 30–60 tons of CO_2. The pores (the stomata) of the succulent plant stay closed during the day to reduce evaporation and open in the cool of the night to absorb carbon dioxide (CO_2). Crassulacean acid metabolism (CAM) allows maguey to absorb CO_2 at night. C_4 acids form separately in time and space. At night PEP carboxylase fixes CO_2 in the cytosol. The malate or other organic C_4 acids are the result of this process and will be transported to the vacuole. Photosynthetic C_3 metabolism is stimulated by daylight in the chloroplasts, fixing CO_2. A large portion of this comes from malate decarboxylation that was stored during the night. The Calvin cycle fixes the released CO_2, and triose phosphate is then converted into starch, using NADHP (Nobel 1998; Taiz and Zeiger 2010). The maguey's metabolism enables it to thrive in areas with minimal rainfall, extreme temperatures, and poor, shallow soils. Meanwhile its relatively shallow roots below the soil surface spread out horizontally, taking in available moisture and nutrients from the topsoil, especially during the rainy season. In addition, the maguey's propagation through hijuelos (baby plants or pups), which can number up to 50 in some varieties, allows it to be a self-reproducing perennial.

Since pre-Hispanic times, the pulque maguey (*Agave salmiana* Otto ex Salm-Dyck) is cultivated, mainly to produce aguamiel, mezcal and fiber (Mora-López et al. 2011). The main product of this specific plant is pulque and it forms a major part of the cuisine, the culture, and the economy of the Mexican high plateau (Tlaxcala, Hidalgo, Puebla and the state of Mexico). However, the cultivated area decreased between 1940 and 2012 with more than 80% (Aguilar et al. 2014), due to the declining reputation of pulque, theft of the leaves and mixiote (leaf peel), the long lifecycle of the plant, and the lack of technology (José-Jacinto and García-Moya 2000; Aguilar et al. 2014; Álvarez et al. 2018). Still, consumption of pulque is very popular (Erlwein et al. 2013), which is celebrated by organizing annual fairs in the Mexican high plateau (Nieto Aquino et al. 2016).

Traditionally, farmers have planted maguey alongside their properties or where erosion must be controlled. Agaves and several of their tree companions have been used as sources of food, beverage, and fiber by indigenous societies for hundreds, in fact thousands of years (Álvarez-Fuentes et al. 2015). The nectar is extracted from the heart of the plant to make pulque. The large leaves, called '*pencas*' which contain up to 30% sugar, would normally be discarded. By regularly pruning a small portion of the leaves, finely chopping them, and fermenting them, farmers could produce inexpensive feed that their animals both enjoyed and thrived on.

Starting in the third year after planting, these crops could produce nearly six tons of animal feed per hectare, without the need of water or fertilizer. As the plants mature, this yield increases and the feed can be used for goats and sheep, and in smaller quantities for pigs and chickens. This cultivation requires minimal maintenance.

The trees produce protein-rich pods that can be ground up and fermented along with the finely chopped agave leaves, all part of a well-rounded agroforestry system. Furthermore, maguey can be used to make inulin, a dietary fiber, and a syrup substituting sugar, and pellets to use in pet food (ECPA 2022).

5.4 Greening Drylands

The dense inter-cropping of high-biomass, high-forage producing, fast-growing maguey species (1600–2000 per hectare) between existing deep-rooted, nitrogen-fixing tree species such as mesquite, or tree seedlings, along with the use of fermented agave/mesquite silage, arid landscapes and livelihoods that are susceptible to droughts can be tremendously transformed (Cummings and Leu 2021). Arid landscapes could become more fertile and productive.

In agroforestry or silvopasture systems, rotational grazing with sheep and goats is practiced, where they are fed on pasture forage and fermented silage, which is also suitable for other livestock, including cattle, pigs, and even chickens. The fermented silage becomes a valuable animal feed by using natural fermentation to turn the indigestible saponin and lectin of the maguey leaves into digestible carbohydrates and fiber. After 30 days the fermented end-product is nutritious and inexpensive. Ten kilos of feed can be converted into one kilo body weight in lambs. The maguey can continuously generate large amounts of biomass, up to 43 tons per hectare (Kant 2010), improving soil fertility, rejuvenating grass and capturing huge amounts of CO_2; The system can be applied anywhere agave grows naturally and is suitable not only in Texas and the Southwestern US, Mexico, Central America, Argentina, Chile, and other Latin American countries, but also in arid and semi-arid regions of North America, Africa, the Middle East, Tunisia, Algeria, India, Pakistan, Afghanistan, Myanmar (Burma), Russia, Hawaii, the West Indies, Puerto Rico, and Australia (Ramos 2007).

5.5 Regenerative Practices

Regenerative and organic agriculture aims to grow crops while at the same time improve and regenerate natural systems. It:

- Increases resiliency in the face of threats such as climate change.
- Improves the soil conditions.
- Prioritize the natural processes such as the water cycle.
- Brings about social and economic benefits for the community.
- Diversifies cultivation.
- Increase the productivity on the long term.
- Capture CO_2.

In Mexico much of the 155 million hectares of crop and pastureland is degraded and no longer suitable for cultivating food crops. On only two to four million hectares of land 37–74% of the country's net current fossil fuel emissions can be sequestered. In countries with less intensive greenhouse gas emissions these could reach a net-zero emission. By implementing this organic agroforestry planning not only the soil and the climate benefits, it also offers marginal farming communities an economic perspective (Cummings and Leu 2021).

5.6 A Maguey Landscape

The abovementioned principles can be used to transform the arid landscape around the twintropolis of Saltillo and Monterrey. The landscape can be adapted in specific areas to cultivate maguey and produce a wide range of products derived from it.

For choosing the location the landscape and ecoregions of northeastern Mexico, such as reflected in the map (Fig. 5.4) have been analyzed. Ecoregions are geographical unities with similar characteristics of flora, fauna, and ecosystems. The environmental and landscape conditions distinguish one from the other. They differ in the functioning of the vegetation in structure and composition of specific species, but also in their physiographical appearances such as mountains, plateaus, plains

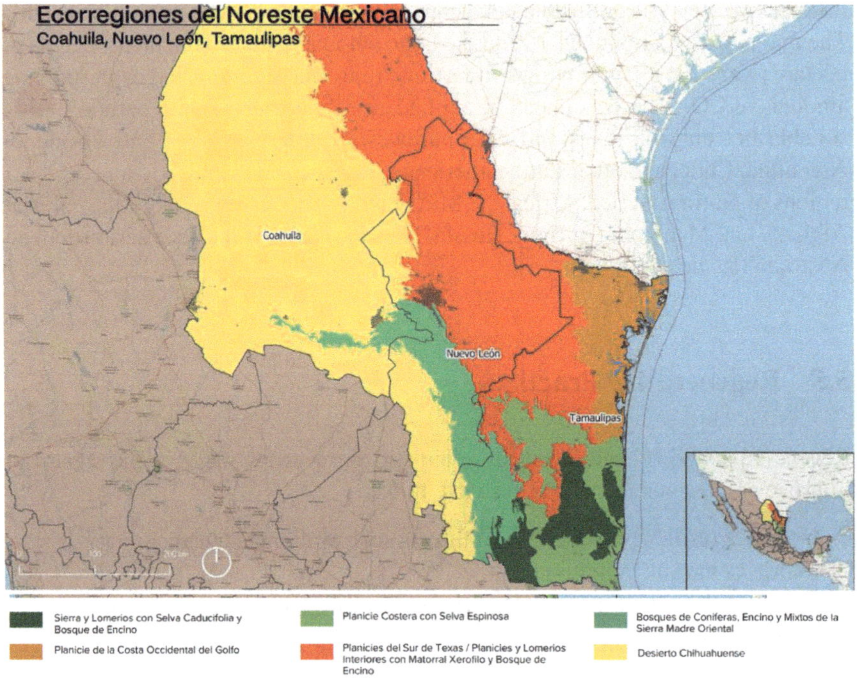

Fig. 5.4 Ecoregions of northeastern Mexico

and watersheds, and their specific climatic conditions such as humidity and temperature. The cultivation of maguey is most effective in drylands; therefore, the desert of the north is selected as the most interesting ecoregion to implement it. Of the six ecoregions, the Chihuahua desert is the most arid.

In this ecosystem, plant species such as pinonero mexicano, chapote, agave mechucdo, oregano, palmito, huizache, mezquite, corona de cristo, and maguey cenizo, dominate the landscape. The 'desierto Chihuahuense' (Fig. 5.5) is approximately 11 million Ha without including the mountainous areas, rivers and waterbodies, roads and highways, and the cities and towns, which are unsuitable for cultivating maguey.

The landscape plan proposes cultivating maguey in scattered clusters throughout the desert (Fig. 5.6). When appr. 2000 maguey plants and 400 mezquite trees are planted per hectare, each hectare will capture 30–60 tons of CO_2 (on average: 45 ton/ha). This system has the potential to capture 500,000 tons of CO_2 across the entire landscape—an enormous amount, exceeding the annual emissions of the entire country.

The landscape plan illustrates the significant contribution this practice could deliver in rural areas. However, in the urban landscape, similar achievements are possible. When the same number of maguey plants and mesquite trees are planted in strips along residential streets (Fig. 5.7), every 4 km could potentially capture the

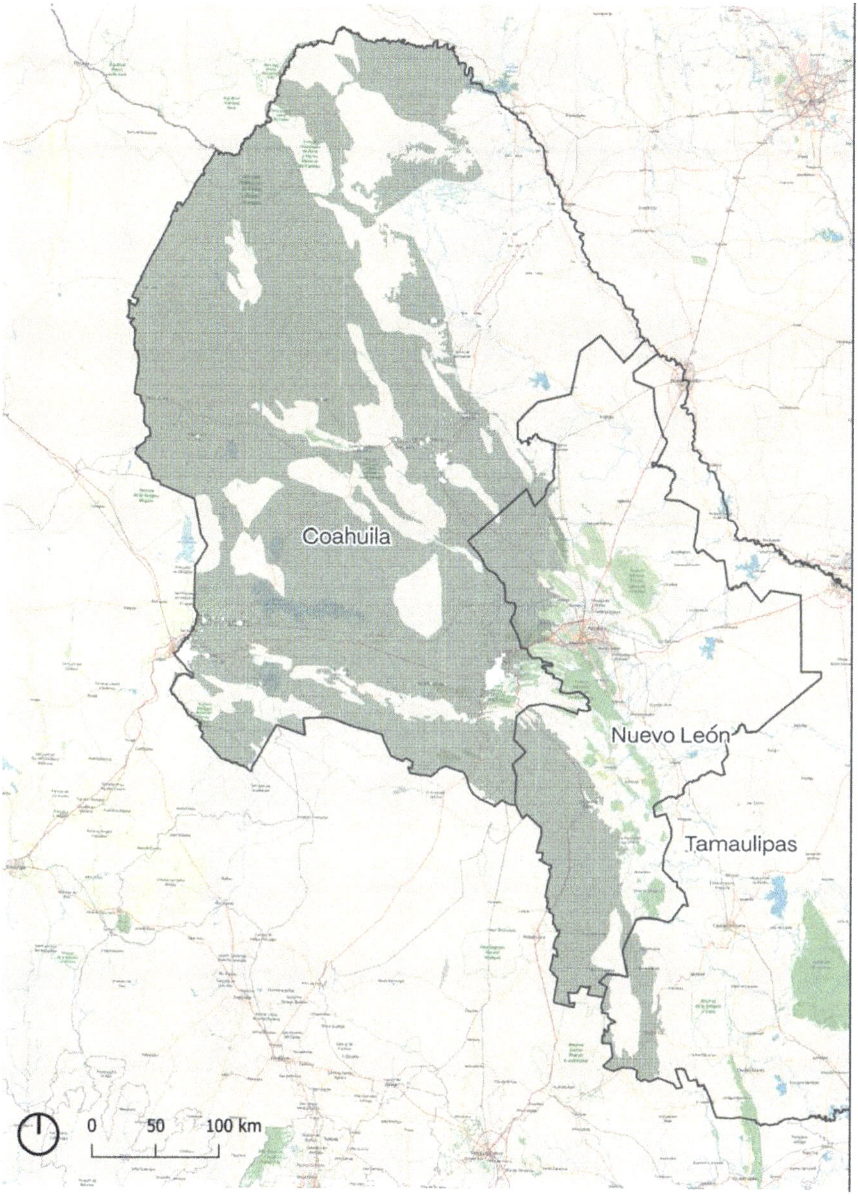

Fig. 5.5 The Chihuahua desert

same amount of CO_2 as 1 ha of land, approximately 45 tons. Specifically, five maguey plants and one mesquite tree would need to be planted every 10 m of road.

At the neighborhood level, such as the Burocratas del Estado area, which spans 44.5 ha, the network of maguey-lined streets would not only capture carbon and

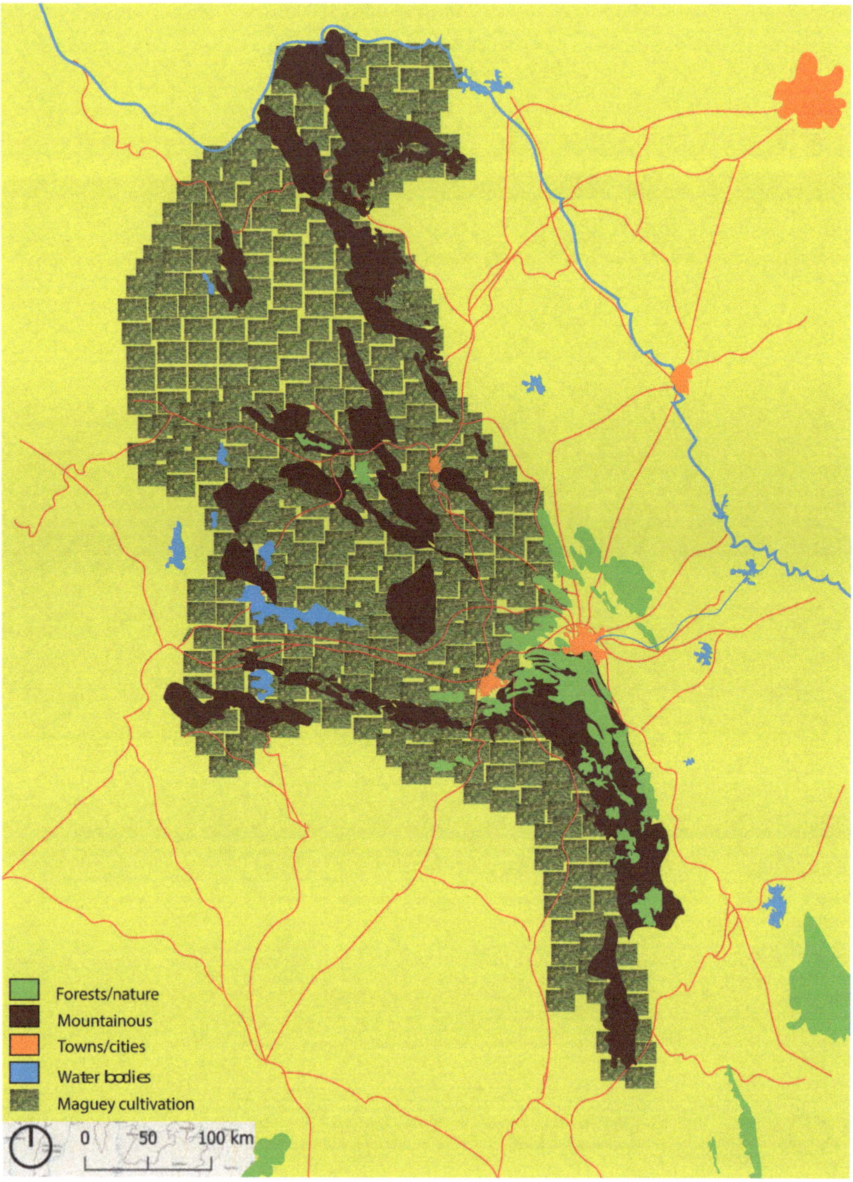

Fig. 5.6 Plan for a Maguey landscape in the Chihuahua desert

enhance street-level livability but also "green" the entire precinct (Fig. 5.8) This helps to lower summer temperatures and decrease biodiversity losses. By planting maguey strips on both sides of every street in the neighborhood, a total of 22,650 m of maguey strips can be implemented, equivalent to capturing 265 tons of CO_2 within this neighborhood alone.

Fig. 5.7 A 'Maguey street'

For the entire metropolitan area of Monterrey (86,960 ha with roughly 5,5 million inhabitants), a rough calculation learns that in total 518,075 tons (86,960/44.5 = 1955 × 265) of CO_2 can be captured by applying this solution. This is, again, more than the entire country of Mexico emits annually.

5.7 The Benefits of Maguey

The development of maguey landscapes has multiple benefits. First and foremost, the carbon it can potentially capture is significant, and could contribute to mitigating climate change. Every hectare planted with maguey could capture 30–60 tons of CO_2 (Cummings and Leu 2021). If this practice is scaled up to arid and semi-arid landscapes, its influence on global climate can be enormous. Together with intact old-growth tropical rainforest this is amongst the highest rates of carbon capture worldwide (Cummings and Leu 2021). In Mexico, the average of 2000 maguey plants planted per hectare could be even higher, up to 3400 plants (Crawford et al. 2022) which would increase the average carbon capture per hectare to 65 tons CO_2.

Besides the beneficial influence on the global climate, cultivating maguey has also positive effects on the landscape itself. As it increases the humidity, it is capable of restoring the soil and bringing the temperatures down.

Fig. 5.8 Greening the Burocratas del Estado neighborhood

The plant's roots help retain water in the soil and revitalize dry earth. This makes it an ideal crop for desert landscapes. It regenerates the landscape and restores the ecology of the desert, thus increasing biodiversity.

Furthermore, the leaves of the maguey, when finely chopped can be fermented and used as animal feed.

The sap of the maguey (aguamiel) can be consumed as a functional, nutritious drink rich in prebiotics or fermented into pulque. This sap offers significant health benefits, being rich in nutrients such as vitamin C, potassium, essential minerals, antioxidants, and amino acids, as well as prebiotics that support healthy gut bacteria. Additionally, the medicinal properties of the plant are notable; the mildly toxic skin of the leaves is claimed to have a positive effect in treating early stages of skin cancer. (Anajwala et al. 2010; Gutierrez et al. 2017).

The plant's fibers are thin yet incredibly strong, making them ideal for creating clothing materials. Additionally, the leaves can be used for roof tiling.

5.8 Conclusion

The maguey plant, along with many other agave species, has the potential to address several of today's global challenges. These plants can capture significant amounts of carbon, and when cultivated on a large scale, they could mitigate up to half of global greenhouse gas emissions. With accelerated climate change, this offers a crucial resource for reducing or even reversing that trend.

The maguey plant excels at processing carbon and extracting it from the air, without competing with urbanized or fertile land uses. It thrives in arid or semi-arid conditions, and therefore the deserts are the most suitable places for its cultivation. It enhances degraded arid soils and reverses desertification, which in turn helps reduce heat buildup and contributes to mitigating global warming. This also halts ecological degradation in these landscapes, allowing for their revitalization.

The plant has several sub-benefits, as it produces healthy products for people, medicine and materials for clothing and construction. The economic potential is also large, as it can help local farmers turn their plants into commercially viable products, and they can potentially capitalize on carbon credits. Global trends in functional foods from naturally sourced ingredients can lead to the development of a range of highly functional and innovative products derived from aguamiel.

The plant offers several additional benefits, producing health-promoting products, medicine, and materials for clothing and construction. Its economic potential is significant, as it can help local farmers turn their crops into commercially viable products while potentially capitalizing on carbon credits. Moreover, global trends in functional foods from naturally sourced ingredients can lead to the development of a variety of highly functional and innovative products derived from aguamiel.

References

Aguilar JB, Enríquez del Valle JR, Rodríguez-Ortiz G, Granados SD, Martínez C (2014) El estado actual de *Agave salmiana* y *A. mapisaga* del Valle de México. Rev Mex Agroecosistemas 1:106–120

Álvarez DMC, García-Moya E, Suárez-Espinosa J, Luna-Cavazos M, Rodríguez-Acosta M (2018) Conocimiento tradicional, cultivo y aprovechamiento del maguey pulquero en los municipios de Puebla y Tlaxcala, México. Polibotánica 45:205–222

Álvarez-Fuentes G, García-López JC, Pinos-Rodríguez JM, Jasso-Pineda Y, Tristán-Patiño FM, González-Garduño R (2015) Maguey (Agave spp.) silage production with either alfalfa or mesquite pod meal as protein sources. J Anim Plant Sci 24(1):3714–3721

Anajwala CC, Patel RM, Dakhara SL, Jariwala JK (2010) In vitro cytotoxicity study of agave americana, strychnos nuxvomica and areca catechu extracts using mcf-7 cell line. J Adv Pharm Technol Res 1(2):245–252. PMID: 22247852; PMCID: PMC3255432

Climate Copernicus (2024) New record daily global average temperature reached in July 2024. Published online 24 July 2024. Available online: https://climate.copernicus.eu/new-record-daily-global-average-temperature-reached-july-2024. Accessed 22 Aug 2024

Crawford W, Tan DKY, Van Ogtrop F (2022) Optimal planting density of *Agave* for maximising aboveground biomass: a systematic literature review. Front Chem Eng 4 (December 2022), Sec. Sustainable Process Engineering. https://doi.org/10.3389/fceng.2022.1039675

Cummings R, Leu A (2021) Best practices: how regenerative and organic agriculture and land use can reverse global warming. Regen Int. Available online: https://regenerationinternational.org/2021/03/08/best-practices-how-regenerative-organic-agriculture-and-land-use-can-reverse-global-warming/. Accessed 23 Aug 2024

Cummins R (2020) Agave power: how a revolutionary agroforestry and grazing system in Mexico can help reverse global warming. Available online: https://regenerationinternational.org/2020/01/20/agave-power-how-a-revolutionary-agroforestry-and-grazing-system-in-mexico-can-help-reverse-global-warming/#:~:text=Agave%20plants%20and%20 nitrogen%2Dfixing,desert%20and%20semi%2Ddesert%20species. Accessed 18 May 2024
De Acosta J (1590) Historia natural y moral de las Indias. Juan de León, Sevilla
ECPA (2022) The billion agave project, 30 November 2022. Available online: https://ecpamericas.org/newsletters/the-billion-agave-project/. Accessed 23 Aug 2024
Erlwein S, Mira Tapia RJ, Velasco Pegueros A (2013) Proceso de elaboración del pulque, su importancia económica y concepción social en Apan, Hidalgo. In: Ramos Ramírez JL (ed) Ejercicios Etnográficos, Aprendiendo a Investigar. ENAH-INAH, Ciudad de México, pp 59–103
Gutierrez JA, Santos-Zea L, Othon SR (2017) Agave syrup extract having anticancer activity. US Patent US9585928B2. Available at: https://patents.google.com/?inventor=Janet+Alejandra+Gutierrez+Uribe&peid=6209e746585e0%3A1ca%3A5f22eacb. Accessed 26 Aug 2024
IUCN. 2024. The IUCN red list of threatened species. Version 2024-1. https://www.iucnredlist.org. Accessed 22 Aug 2024
José-Jacinto R, García-Moya E (2000) Remoción cuticular ("Mixiote") y desarrollo foliar en los Agaves Pulqueros (*Agave salmiana* y *A. mapisaga*). Bol Soc Bot Méx 66:73–79
Kant P (2010) Could agave be the species of Choic3 for climate change mitigation? IGREC working paper IGREC-11: 2010, Institute of Green Economy, New Delhi
Mora-López JL, Reyes-Agüero JA, Flores-Flores JL, Peña-Valdivia CB, Aguirre-Rivera JR (2011) Variación morfológica y humanización de la sección Salmianae del género Agave. Agrociencia 45:465–477
Nieto Aquino R, Vargas Monter J Nieto Aquino JC, Rodríguez Ortega A, Jiménez Pérez VM, Callejas Hernández J, Ortiz Balderas M (2016) El cultivo de maguey pulquero (*Agave salmiana*) en el valle del mezquital. Universidad Politécnica de Francisco I. Madero, Hidalgo. 52 p
Nobel PS (1998) Los Incomparables Agaves y Cactos. Editorial Trillas, México. 211 p
Ramos MG (2007) The Ubiquitous Mesquite. The Texas Almanac Available online: https://www.texasalmanac.com/articles/the-ubiquitous-mesquite. Accessed 23 Aug 2024
Semarnat y CP (2003) Evaluacion de la degradacion del suelo causada por el hombre en la Republica Mexicana, escala 1:250 000
Taiz L, Zeiger E (2010) Chapter 8: photosynthesis. In: Plant physiology, 5th edn. Sinauer Associates Inc., Publishers, Sunderland, pp 200–235
Turok M (2010) El Maguey símbolo de identidad nacional. Available online: http://jorgeobregon.com.mx/wp/wp-content/uploads/2012/05/maguey.pdf. Accessed 22 Aug 2024
United Nations Convention to Combat Desertification (2019) The global land outlook. Latin America and the Caribbean Thematic Report, Bonn
Von Humboldt A (1827) Ensaya político del reino de la Nueva España. Porrúa, Mexico City, p 281
WMO (2024) State of the global climate 2023. WMO report no. 1347. Geneva, Switzerland. Available online: https://library.wmo.int/viewer/68835/download?file=1347_Global-statement-2023_en.pdf&type=pdf&navigator=1. Accessed 22 Aug 2024

Chapter 6
Dry Urbanism Solutions for Arid Futures

Lizbeth Anahi Saavedra Flores and Jorge Gerini Igartua

Abstract Climate change intensifies the challenges faced by urban areas in arid regions, necessitating the development of innovative and sustainable solutions. This research explores the concept of dry urbanism as a framework to address these challenges, focusing on the role of design and interventions in both hyper-arid regions and urban settings. The approach seeks to resolve existing issues, such as resource scarcity in cities, while also functioning as a proactive strategy for building resilience by understanding and addressing the specific needs of arid urban environments, both in the present and the future.

A case study of the Master Plan for Ciudad Juárez was conducted to identify key strategies for water management, urban design, and community engagement. The findings emphasize the critical importance of equitable resource distribution, green infrastructure, and resilient urban planning in the creation of livable and sustainable cities in arid regions. Furthermore, the Polanco Cultural Corridor case study illustrates the practical application of dry urbanism principles, demonstrating the potential to transform urban spaces into resilient, people-centered environments. By integrating these strategies into urban planning, cities can enhance their adaptability to climate change, improve the quality of life, and promote sustainable development.

Keywords Climate change · Urban areas · Arid regions · Dry urbanism · Sustainable solutions · Water management · Urban design · Resilience · Resource scarcity · Green infrastructure · Master plan · Community engagement · Livable cities · Sustainable development

---

L. A. Saavedra Flores (✉) · J. Gerini Igartua
Sordo Madaleno Arquitectos, Mexico City, Mexico

R. Roggema (ed.), *Dry Urbanism*, Contemporary Urban Design Thinking,
https://doi.org/10.1007/978-3-031-77347-1_6

6.1 Climate Change and Urban Adaptation in Arid Environments

Climate change presents formidable challenges for urban areas, particularly those located in arid environments. These challenges manifest in heightened droughts and extreme temperatures, threatening both residents' quality of life and the sustainability of urban infrastructure (IPCC 2021). In response to these realities, it becomes imperative to develop robust adaptation strategies that effectively confront and mitigate these impacts. Essential measures include efficient water management and the promotion of resilience in urban infrastructure, as integral components of sustainable urban design, often referred to as "dry urbanism."

Effective water management stands out as a critical challenge in arid regions facing climate change. The scarcity of water necessitates sustainable policies ensuring its continuous supply and conservation. As emphasized by the World Bank, effective water management is pivotal for sustainable development and climate resilience in water-scarce regions (Zhang and Borja-Vega 2024). Strategies such as rainwater harvesting offer a promising solution by capturing and storing rainwater for irrigation and non-potable uses. Additionally, recycling greywater from sinks, showers, and washing machines is crucial for reducing demand on potable water supplies. These practices must be seamlessly integrated into urban planning to promote efficient and sustainable water usage across agricultural and urban landscapes.

Sustainable urban design plays a crucial role in minimizing cities' ecological footprint and enhancing energy efficiency, particularly under the paradigm of dry urbanism. Designing buildings with passive cooling systems, leveraging natural ventilation, and using construction materials that reduce reliance on artificial air conditioning are fundamental strategies. Passive cooling designs significantly lower energy consumption while improving indoor comfort. These buildings not only consume less energy but also enhance residents' thermal comfort (Taleb 2014). Furthermore, promoting public transport and non-motorized modes like cycling is essential for reducing greenhouse gas emissions and alleviating urban infrastructure burdens.

The integration of green spaces is another vital aspect of sustainable urban design in arid cities. These spaces not only enhance air quality and urban biodiversity but also mitigate the urban heat island effect, where urban areas become significantly warmer than their rural surroundings due to high building and pavement density. Urban green spaces can reduce local temperatures by up to 2 °C, offering significant relief in hot climates (Farshid Aram 2019). Parks and community gardens serve as carbon sinks, provide recreational areas for residents, and enhance urban quality of life.

Food security in arid environments faces increasing threats from climate change, underscoring the importance of enhancing cities' capacity to produce and distribute food sustainably. Urban agriculture emerges as a key strategy, enabling food cultivation in available urban spaces such as rooftops, walls, and vacant lots. These practices improve access to fresh, nutritious food and reduce dependence on external supply chains vulnerable to extreme weather events. The Food and Agriculture Organization highlights urban agriculture's potential to bolster food security and nutrition, especially in rapidly urbanizing regions (FAO 2022).

Community participation and education are pivotal for the success of climate change adaptation strategies in urban settings. Engaging communities in urban development planning ensures solutions that are inclusive and responsive to local needs. The United Nations Development Program underscores community engagement as essential for building resilience and achieving sustainable urban development (UN 2022). Public education campaigns on climate change impacts and actions for mitigation empower citizens to adopt sustainable practices and support community initiatives that bolster urban resilience.

After these considerations, we have concluded that adapting to climate change in arid urban environments necessitates comprehensive strategies encompassing water management, sustainable urban design, food security, and community engagement. Dry urbanism provides a cohesive framework for integrating these strategies, promoting resilient, efficient, and sustainable urban development. Collaborative efforts involving all stakeholders are essential to building cities capable of confronting climate challenges and ensuring a sustainable future for residents.

6.2 Equitable Urban Design in Arid Environments

Urban design in arid environments holds a significant promise as a catalyst for advancing social justice. In these regions, equitable resource distribution, inclusion of marginalized communities, and resilience to climate-related challenges are paramount. Thoughtful urban design can address these issues to foster greater social equality and cohesion among residents.

Ensuring fair access to essential resources such as water, housing, and energy is a cornerstone of urban design in arid environments. Effective urban planning can distribute these resources equitably across all sectors of the population, thereby reducing disparities and promoting social equality. The United Nations underscores that "equitable access to resources is crucial for alleviating urban poverty and fostering sustainable development" (UN 2022). For example, implementing sustainable water management systems like rainwater harvesting and greywater reuse can ensure reliable water supply to all neighborhoods, irrespective of their socio-economic status. Similarly, energy-efficient housing designs can lower utility costs and improve living conditions for residents, including those from economically disadvantaged backgrounds.

By creating inclusive spaces that cater to their specific needs, urban design can enhance social integration and encourage active participation in urban life. This may involve developing affordable housing projects, community centers, and public spaces that are accessible and welcoming to all citizens. According to the World Bank, "inclusive urban planning ensures that all residents, particularly the poor and marginalized, benefit from urban development" (World Bank 2023). Prioritizing the voices and needs of marginalized communities in planning processes can foster a more inclusive and equitable urban environment.

Urban design also plays a pivotal role in bolstering resilience to climate-related crises, which pose specific challenges in arid environments. Cities in these regions confront unique risks such as severe droughts and extreme temperatures, disproportionately affecting vulnerable populations. Integrating adaptation and mitigation measures into urban design can safeguard these communities and enhance their resilience. For instance, designing buildings with passive cooling systems and using materials that minimize heat absorption can help maintain comfortable indoor temperatures without excessive energy consumption. Additionally, creating green spaces can mitigate the urban heat island effect and provide venues for recreation and community gatherings. As emphasized by the UN, "adaptive urban design is critical for enhancing cities' resilience to climate impacts" (UN 2022).

In conclusion, urban design for arid cities can serve as a potent instrument for promoting social justice by ensuring equitable resource access, embracing marginalized communities, bolstering climate resilience, and fostering active citizen participation. By prioritizing these principles, urban planners can shape cities that are not only sustainable and resilient but also inclusive and equitable, ultimately enhancing the well-being and unity of all residents.

6.3 Practical Experience: Master Plan Ciudad Juárez

The Master Plan Ciudad Juárez (MPCJ) stands out as an exemplary model of urban design that prioritizes climatic conditions. This project, designed by Sordo Madaleno Arquitectos, demonstrates the feasibility of creating sustainable, livable, and resilient cities capable of adapting to the specific needs of their environment. By emphasizing thermal comfort, protection against the elements, and the creation of high-quality public spaces, the MPCJ promotes the well-being and quality of life of Ciudad Juárez inhabitants, transforming the city into a more prosperous and enjoyable place to live.

6.3.1 Distinct District and Diverse Experiences

The development is structured into four Distinct Districts (Fig. 6.1), each with its own unique identity and diverse experience:

- Lake District: Spanning an area of 46,906 m^2, this district is characterized by its urban structure centered around a 9796 m^2 lake. This configuration creates a dynamic space where commercial and residential uses converge in a natural environment.

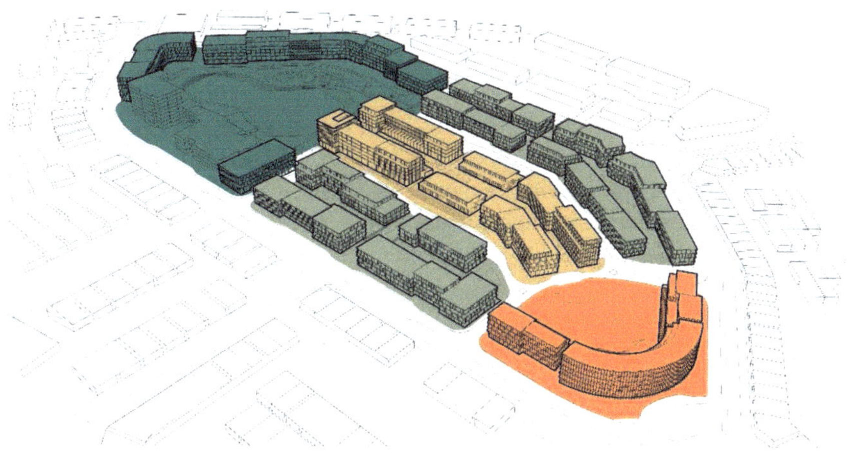

Fig. 6.1 Districts' proposal

- Commercial District: Covering 17,827 m^2, this area accommodates mixed-use developments and features a central pedestrian axis lined with commercial spaces. A market is located at the intersection of the pedestrian axes within the development.
- Residential District: Encompassing 29,586 m^2, this district is divided into two areas adjacent to the commercial district. Residential areas feature enclosed courtyards that foster family dynamics and recreation within a secure environment that facilitates contact with nature.
- Park District: This recreational space includes an amphitheater next to a smaller lake. Public space interventions promote interaction between commercial areas and mixed-use ground-floor commercial spaces. The total area of the district is 10,424 m^2.

Despite the density of the master plan, various strategies are implemented to achieve comfort and enjoyment of both outdoor and indoor spaces in buildings. It is crucial to note that the vision extends beyond the incorporation of specific architectural elements; from the project's conception, there is an intentional effort to create a sustainable and legible design that, through its form, orientation, and layout, contributes to this objective.

The MPCJ presents three main contributions:

1. Demonstration of the feasibility of climate-centric urban design: This project encases how cities can adapt to their climatic context and create habitable and sustainable spaces.
2. Creation of high-quality public spaces: The districts of the MPCJ prioritize the generation of public spaces that promote social interaction, well-being, and recreation.

3. Focus on habitability and quality of life: The design of the MPCJ focuses on improving the quality of life for Ciudad Juárez inhabitants, creating a more prosperous and pleasant living environment.

The Master Plan Ciudad Juárez emerges as an innovative model of climate-centric urban design, offering a replicable approach for the creation of sustainable, livable, and resilient cities that respond to the specific needs of their context. This project represents a significant step towards building a more prosperous and sustainable future for cities worldwide.

6.3.2 The Shape of City Blocks: A Strategic Approach to Microclimate Creation

In urban design, the shape of city blocks plays a crucial role in creating favorable microclimates for thermal comfort and livability. The proposed "typical" city block, with two rows of buildings on the north and south ends and a central green corridor, serves as an effective strategy to mitigate the impact of climate and generate more pleasant public spaces (Fig. 6.2).

There are three main strategies that have been followed to achieve this purpose, these are explained as follow:

1. The arrangement of buildings at the north and south ends of the block creates a natural barrier against direct solar radiation, especially during peak sunlight

Fig. 6.2 Master Plan Ciudad Juárez

hours. This results in a significant reduction in the surface temperature of the ground and air within the block, creating shaded and cooler areas.

2. The central green corridor functions as an urban oasis, where the presence of vegetation increases ambient humidity through evapotranspiration. This natural cooling process, combined with the shade provided by buildings, creates a cooler and more pleasant microclimate compared to surrounding areas.
3. The elongated shape of the block, with a central green corridor, promotes natural air circulation. The wind flows freely through the open space, creating a fresh breeze that renews indoor air and dissipates accumulated heat.

The combination of these factors—urban heat mitigation, oasis effect, and natural ventilation—significantly improves thermal comfort within the block. Users can enjoy cooler and more pleasant public spaces, especially during warm climates. The effectiveness of this strategy depends on various factors, such as block orientation, building height, vegetation density, and local climate characteristics.

6.3.3 Green Corridor Network

In urban design, the strategic integration of greenery has become a fundamental tool to mitigate urban heat impact and create favorable microclimates for thermal comfort and livability. Studies have shown that plants can decrease site temperatures by up to 4 °C through mechanisms such as shading from leaves, evapotranspiration cooling, and albedo effect reflecting solar radiation, thereby reducing heat absorption by the ground (Taleb 2014).

Aware of this potential, the project incorporates central green corridors that connect to form a network of urban oases. This strategy aims not only to cool the city and create more pleasant microclimates but also to achieve broader social and environmental goals (Fig. 6.3).

Fig. 6.3 Landscape strategy

Vegetation plays a pivotal role in urban environments by providing shade and augmenting evapotranspiration processes, thereby creating cooler and more comfortable public spaces, particularly advantageous in warmer climates. This enhancement significantly improves the quality of life for urban inhabitants, enabling them to enjoy outdoor settings without discomfort from excessive heat.

Additionally, plants function as natural filters, absorbing carbon dioxide (CO_2) and emitting oxygen, thereby enhancing air quality. This ecological service is especially critical in urban areas where high levels of air pollution pose significant health risks and environmental challenges.

Moreover, green spaces promote and encourage outdoor recreational activities, fostering an active and healthy lifestyle among urban residents. Green corridors, specifically designed for activities such as walking, jogging, and cycling, positively contribute to physical fitness and mental well-being. Green corridors serve as communal gathering spots that facilitate social interaction and community cohesion among residents. These public spaces play a pivotal role in strengthening social bonds, enhancing neighborhood solidarity, and cultivating a shared sense of belonging.

The integration of green corridors in the Master Plan is not only a strategy to mitigate urban heat but also an opportunity to create more pleasant public spaces, promote physical activity, social interaction, and strengthen community cohesion (Fig. 6.4). It exemplifies how urban design can contribute to the creation of more sustainable, healthy, and vibrant cities. Lastly, vegetation within urban landscapes attracts diverse urban wildlife, thereby fostering a more balanced and ecologically diverse ecosystem. This supports conservation efforts aimed at preserving local

Fig. 6.4 Internal green corridor

biodiversity while enhancing the aesthetic appeal and natural ambiance of urban environments.

Moreover, all the vegetation incorporated into these green corridors is native to the region, drawing inspiration from the local landscape (Fig. 6.5). This deliberate choice not only ensures the plants' resilience and adaptability to the existing climatic conditions but also reinforces the cultural and ecological identity of the area. By mirroring the indigenous flora, the design pays homage to the natural environment, creating a seamless integration between the urban and the natural world. This approach not only preserves the local ecosystem but also educates and connects the community to the natural heritage of their region, further enriching the urban experience.

Fig. 6.5 Proposed vegetation

6.3.4 Volumetric Stepping and Double Facades

In this project, there are two main strategies to achieve both the enhancement of user experience while inhabiting both outdoor and indoor spaces, looking for the creation of harmonious, comfortable urban spaces at a human scale; these are volumetric stepping and double facades.

The proposal for volumetric stepping (Fig. 6.6) aims to create a gradual transition in building height, better integrating the urban image with the surroundings and allowing natural light to enter streets. Setbacks from the third and fourth levels allow for the incorporation of vegetation, terraces, and balconies, adding aesthetic value and generating public spaces for the enjoyment of residents. The proportional setback system per level avoids monotony and adds visual richness to the urban ensemble, allowing greater flexibility in design and better adaptation to the site.

Double facades create an intermediate space between the exterior and interior of the building, acting as a threshold that dissipates heat and regulates temperature, enabling better natural ventilation and reducing the need for air conditioning (Fig. 6.7). Recessed windows or balconies in double facades block direct solar radiation, preventing overheating of the interior of the building and contributing to energy efficiency and thermal comfort for users. The implementation of double

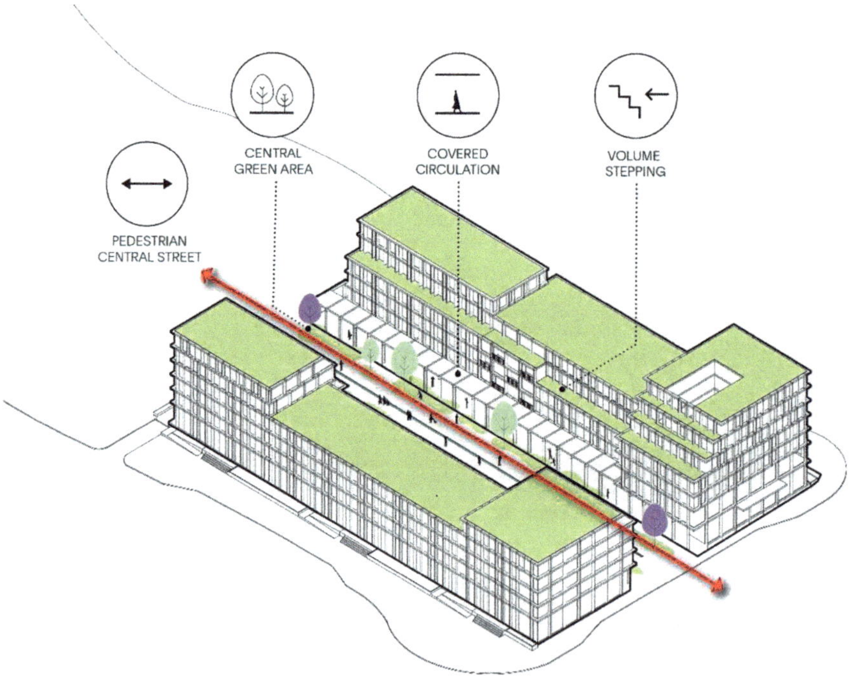

Fig. 6.6 Typical block proposal

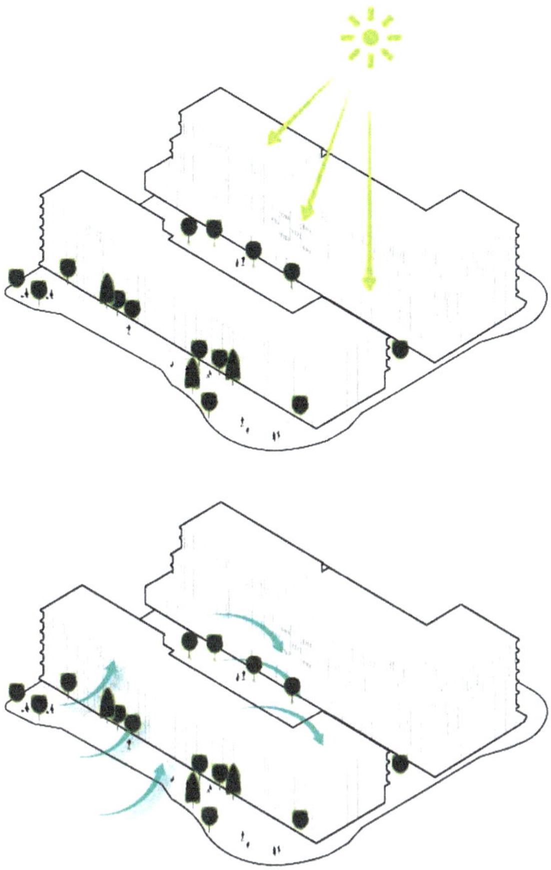

Fig. 6.7 Sunlighting and wind strategies

facades optimizes energy consumption and CO_2 emissions, being a relevant strategy in the fight against climate change.

Volumetric stepping and double facades represent an innovative and responsible approach to urban design, promoting the well-being of citizens and creating more resilient and sustainable cities. These strategies not only shape an attractive and human-scale urban landscape but also contribute to the thermal comfort of buildings, optimizing their energy efficiency and creating a healthier environment.

In environments characterized by extreme and challenging climatic conditions, the importance of designing urban spaces that prioritize shaded pathways cannot be overstated. Ensuring that pedestrians can move comfortably in such areas is essential for fostering a livable and sustainable urban experience. This project emphasizes the critical role of mobility in achieving these objectives by implementing the concept of superblocks—large urban units where vehicular traffic is restricted to the perimeter, thereby creating a pedestrian-friendly environment within (Fig. 6.8).

The strategic design of these superblocks not only enhances pedestrian safety and comfort but also serves as a foundation for integrating green infrastructure into

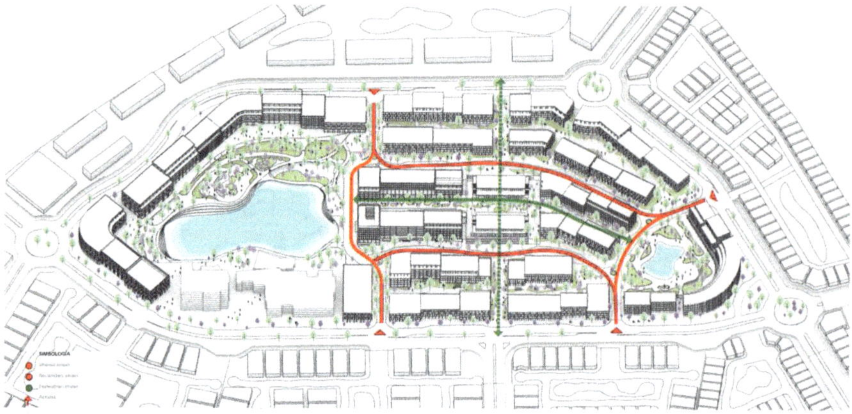

Fig. 6.8 Mobility strategy

urban spaces. By eliminating internal vehicular traffic, the interiors of these blocks are transformed into verdant green corridors. These corridors are carefully curated with native plant species that are specifically chosen for their adaptability to the local climate, thus ensuring ecological sustainability.

This approach not only mitigates the harsh effects of the climate by providing natural cooling through shading but also contributes to biodiversity, enhances the aesthetic value of the urban landscape, and promotes environmental resilience. The project, therefore, represents a holistic approach to urban design, where mobility, environmental sustainability, and human comfort are seamlessly integrated.

6.3.5 Urban Collection and Recycling Network

In the arid urban landscape, where water is a precious resource, the urban collection and recycling network emerges as a fundamental complementary strategy for efficient water management and the creation of more sustainable cities. This network, composed of systems for capturing, storing, treating, and reusing rainwater and greywater, becomes an invaluable ally in mitigating water stress and increasing water availability for various uses.

Its integration with urban design is crucial. The arrangement of buildings, the configuration of streets and public spaces, and the selection of permeable materials optimize the capture and storage of rainwater. Green areas and ecological corridors contribute to aquifer recharge and the creation of cooler microclimates.

The benefits are multiple and can be summarized as follows:

1. **Reduction in potable water consumption:** Collected and recycled water is used for watering green areas, washing streets, replenishing ponds and fountains, and dual flush toilet systems, reducing pressure on freshwater resources.

2. **Improved water quality:** Adequate treatment systems ensure that collected and recycled water meets necessary standards for use. This resource is maximized, and the environmental impacts of its untreated discharge are minimized.
3. **Flood reduction:** Rainwater capture and storage control water flow during torrential rains, reducing the risk of flooding and its negative consequences on urban infrastructure and resident safety.

In arid environments, the selection of appropriate technologies, community education and awareness, and a strong regulatory and financial framework are essential for the success of this strategy. The urban collection and recycling network, integrated with a sustainable urban design approach, represents an innovative response to efficient water management in arid environments, contributing to the mitigation of water stress, improvement of water availability, reduction of environmental impact, and promotion of a responsible water culture among citizens.

This project specifically addresses these necessities by creating an on-site water reservoir with the capacity to support all green areas within the development (Fig. 6.9). This reservoir is part of a broader, water collection and distribution network that spans the entire complex. Rainwater and greywater are systematically captured, treated, and stored, ensuring a consistent supply of non-potable water for irrigation, landscape maintenance, and other uses (Fig. 6.10). This integrated system not only reduces reliance on external water sources but also enhances the resilience of the complex against water scarcity.

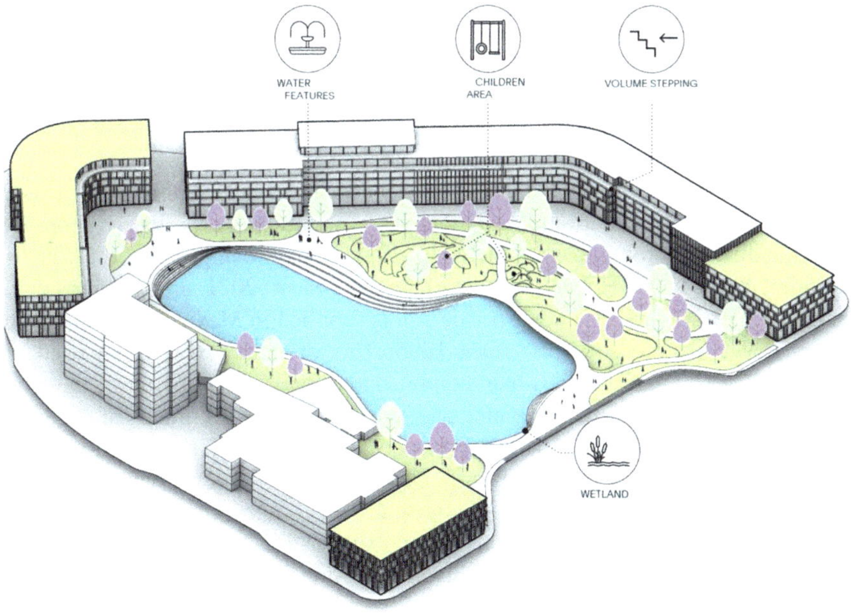

Fig. 6.9 Lake district

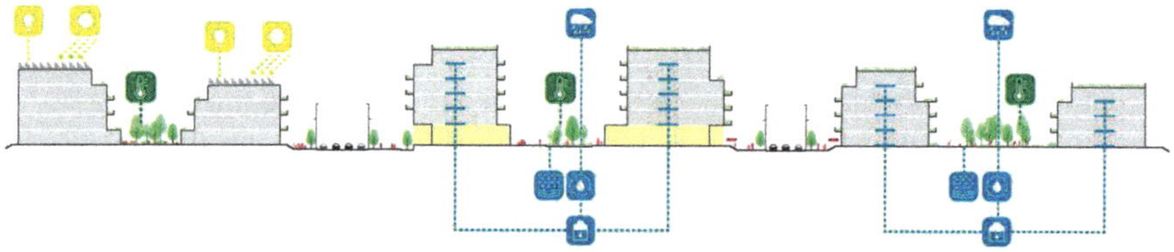

Fig. 6.10 Sustainable strategy for water collection and energy generation

6.4 Responding to Dry Urbanism in Cities: Polanco Cultural Corridor

The emerging trend of "dry urbanism" addresses the challenges posed by climate change and water scarcity, emphasizing the need for cities to adapt to increasingly dry conditions through sustainable urban planning, water conservation, and resilient infrastructure. However, the response to dry urbanism should not be confined to traditionally arid or semi-arid environments; it requires a broader and more proactive approach that encompasses all cities, regardless of their current climate.

Climate change is disrupting global precipitation patterns, potentially transforming areas that currently do not face drought issues into vulnerable regions. Even cities in temperate or humid climates may experience prolonged or intensified drought periods (IPCC 2021). Therefore, anticipating these potential shifts is crucial for ensuring long-term urban resilience.

As cities continue to expand, the demand for water increases exponentially. In many cases, existing water infrastructure is already strained, and a lack of planning could lead to water crises even in historically water-abundant areas (WRI 2019; UN 2018). Implementing dry urbanism principles—such as water efficiency, rainwater harvesting, and greywater reuse—is essential for all cities, not just those currently experiencing arid conditions.

In recent years, Mexico City has faced increasing challenges related to water scarcity, with several alarming developments. The city has experienced more frequent and severe droughts, with the 2020–2021 drought being one of the worst in decades. According to El Economista (2024) at the end of 2021, it was observed that an average of 308 liters of water per person is supplied in Mexico City. This is the lowest level recorded, at least since 1996. This drought severely impacted the Cutzamala System, a critical water supply source for about a quarter of the city's population, leading to water rationing and reduced access for millions of residents.

Moreover, water scarcity has exacerbated social inequalities, particularly affecting marginalized communities. Over the past 5 years, many low-income neighborhoods have endured prolonged periods without reliable access to water, often receiving it only once a week or relying on expensive water trucks. This situation has heightened health risks, especially during the COVID-19 pandemic, when proper hygiene was crucial. (CNN 2024).

This case study examines the "Polanco Cultural Corridor," an urban initiative led by Sordo Madaleno Arquitectos. The project was initiated to address the need for a mitigation plan due to construction activities in the Polanco neighborhood. Following discussions with the relevant authorities, a "green spine" project was developed (Fig. 6.11), designed to connect with other green spaces in the capital. This initiative also aims to be a leading example in the creation of a network of green corridors across Mexico City. This corridor is designed to connect two prominent cultural hubs in Mexico City: the Chapultepec Park and the cultural pole that houses the Jumex Museum and the Soumaya Museum. These hubs represent a substantial portion of the city's cultural assets, drawing significant tourist traffic annually.

Polanco, a neighborhood renowned for its diverse gastronomic, commercial, and cultural offerings, provides an opportunity to enhance both the tourist experience and the quality of life for residents. The concept development is based on the principles outlined below:

1. **Walking as a fundamental human experience:** The project emphasizes providing opportunities for both residents and visitors to navigate the interconnected areas on foot. Walking is a quintessential human activity, encompassing functional, social, and aesthetic dimensions. As a functional experience, walking facilitates movement from one location to another. As a social experience, it enables interaction with others, while as an aesthetic experience, it allows individuals to engage with and appreciate their surroundings. Promoting pedestrian

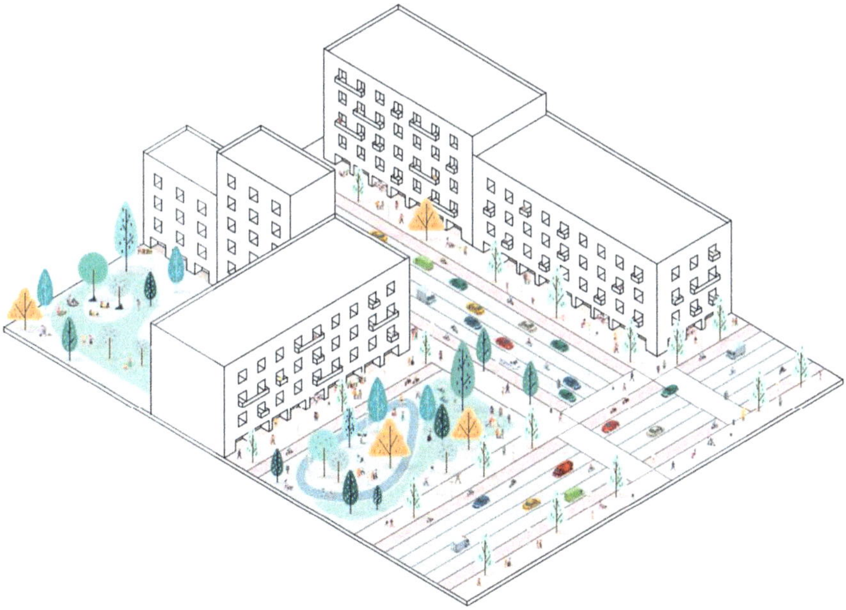

Fig. 6.11 The street we want

activity not only enhances the enjoyment of strolling and social interaction but also redefines the urban environment as a place where residents develop a sense of belonging. The project's integration of traffic reorganization, bicycle lanes, and green spaces reflects a commitment to reimagining the city as a more sustainable, livable, and community-centric space.

2. **Street as public space:** For a street to function effectively as a public space, it must accommodate diverse activities, including pedestrian movement, rest, and leisure. The Paseo Polanco project exemplifies this principle through a focus on landscape infrastructure, smart mobility, and public space design. By integrating these elements, the project aims to create an engaging, secure, and aesthetically pleasing urban environment while preserving the street's essential transportation role. This balanced approach fosters connectivity, economic vitality, and environmental sustainability.
3. **Public space as a vital urban asset:** Public spaces are pivotal to a city's social and economic fabric. Paseo Polanco envisions a network of interconnected plazas, green areas, and pocket parks to create a legible and accessible public realm. These spaces are designed to support a variety of activities, from social gatherings to recreation, enhancing the neighborhood's vitality and fostering a strong sense of community.
4. **Landscape and green infrastructure:** Landscape is integral to Paseo Polanco's urban infrastructure. By establishing extensive green corridors, the project enhances ecological health and visual appeal. These green spaces offer multiple benefits, including carbon sequestration, stormwater management, and improved air quality. Landscape elements are strategically employed to create a legible and comfortable urban environment
5. **Mobility:** The mobility strategy prioritizes pedestrian experience by reconfiguring the street hierarchy. This entails reorganizing traffic patterns, expanding pedestrian infrastructure, and introducing dedicated cycling lanes. The objective is to diminish car dependency and cultivate sustainable transportation options. By incorporating wider sidewalks, optimized street lighting, and tree-lined pathways, the project creates a pedestrian-friendly environment that encourages walking and cycling, fostering a more interconnected and accessible local mobility network.

6.5 The Corridor Design

Although the total length of the corridor is 5 km, the project proposes to achieve this through the development of four distinct corridors aimed at improving mobility, facilitating access to public cultural experiences, and significantly increasing green spaces. Beyond serving as an urban intervention, the Paseo Polanco initiative seeks to transform the designated public spaces into crucial infrastructural elements of the city. The goal is to create environments that promote community engagement and

Fig. 6.12 Corridor Masterplan

vibrant street life, thereby enhancing cultural and commercial exchanges in the area (Fig. 6.12).

6.5.1 First Section: Cultural Corridor

The first segment of Paseo Polanco, the Cultural Corridor, stretches along Ferrocarril de Cuernavaca and Cervantes Saavedra, where significant cultural institutions like the Soumaya Museum, Telcel Theater, Jumex Museum, Inbursa Aquarium, are located. This corridor seamlessly weaves the cultural ambiance of these landmarks into the urban fabric, fostering an immersive blend of art and cityscape. (Fig. 6.13). Streets are redesigned to function as plazas with slow vehicle traffic, prioritizing pedestrian use and allowing the entire street to transform into a recreational space on certain days.

The reorganization of traffic along Av. Ferrocarril de Cuernavaca, connecting with Av. Cervantes Saavedra, creates a plaza that links the Inbursa Aquarium with Antara II. This area, characterized by high pedestrian traffic, is reimagined to offer a spacious, pedestrian-friendly crossing, with vegetation enhancing the walkability and legibility of the space (Fig. 6.14). The strategy also aims to improve connections with Plaza Carso, reducing noise pollution and traffic accidents, and continuing the network of cycle paths to encourage alternative mobility.

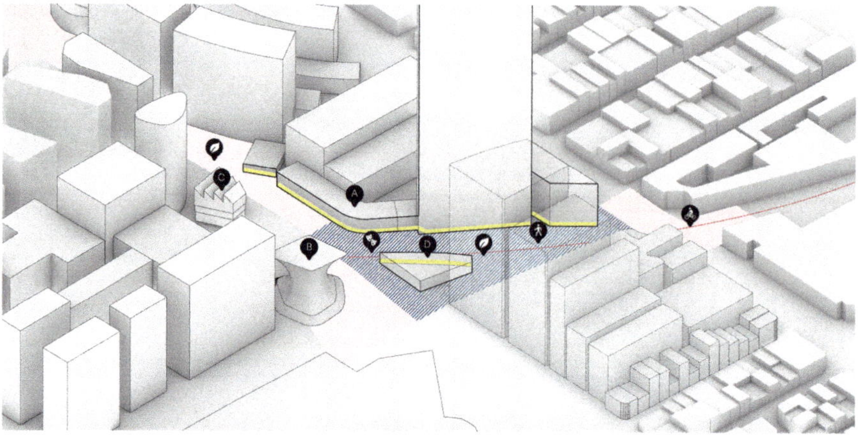

Fig. 6.13 Section 1 strategy A

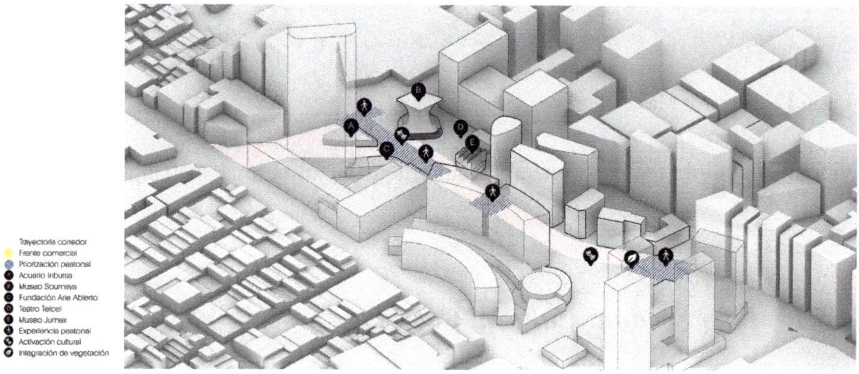

Fig. 6.14 Section 1 strategy B

Cervantes Saavedra, currently marked by chaotic traffic and poor pedestrian infrastructure, will see the implementation of safe crossings and widened sidewalks. The addition of artistic installations and rest areas separated from the roadway by a vegetative strip maintains a rhythm of pink Guayacan trees, further establishing this avenue as the cultural street of the project. The central median will also be expanded to accommodate new plant species, enhancing the street's green infrastructure.

6.5.2 Second Section: Commercial Corridor

The second segment, the commercial corridor, significantly improves mobility in the area by prioritizing walkability and alternative transportation. Sidewalks are widened, and a cycle path is integrated, while the removal of parking lanes

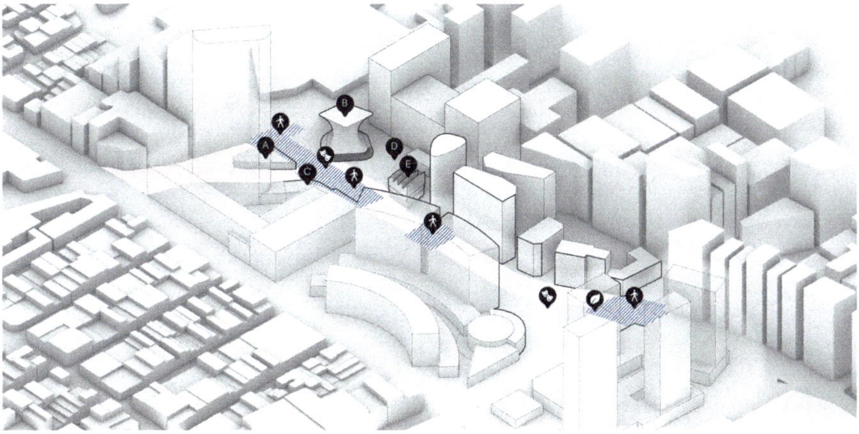

Fig. 6.15 Section 2 strategy

redistributes space without reducing functionality for vehicles. This not only enhances the transit experience but also creates a comfortable and safe commercial promenade, protected by greenery and offering shaded rest areas under tree canopies.

Along Av. Moliere, resting spots and cultural activation areas are established, ensuring the presence of art installations and pavilions throughout the corridor (Fig. 6.15). The dual commercial and residential nature of this avenue is addressed by providing spaces for loading and unloading goods without disrupting traffic, improving pedestrian infrastructure, and integrating vegetation along the avenue.

6.5.3 *Third Section: Linear Park*

The third segment of Paseo Polanco focuses on the linear park, which spans Emilio Castelar, Virgilio, and Julio Verne streets around Parque Lincoln. This area, surrounded by restaurants, luxury shops, and other attractions, generates a distinct atmosphere within the project, drawing large crowds. The intervention here is subtle, focusing on pavement changes, vegetation accents, and the creation of plazas in underutilized spaces (Fig. 6.16). Virgilio Street is proposed as a nearly pedestrian-only area, creating a pleasant corridor with abundant greenery, enhancing the visitor experience to local restaurants and shops.

The identity of Parque Lincoln as a vital green space is preserved and emphasized by improving sidewalks and pedestrianizing Virgilio Street. Julio Verne Street is also calmed to create a vibrant, local, and enjoyable area. Access to this landmark will be facilitated by improved sidewalks and the creation of "complete streets" along Av. Emilio Castelar, particularly between Av. Presidente Masaryk and Edgar Allan Poe, with green borders separating pedestrian spaces from the roadway and safe crossings at intersections.

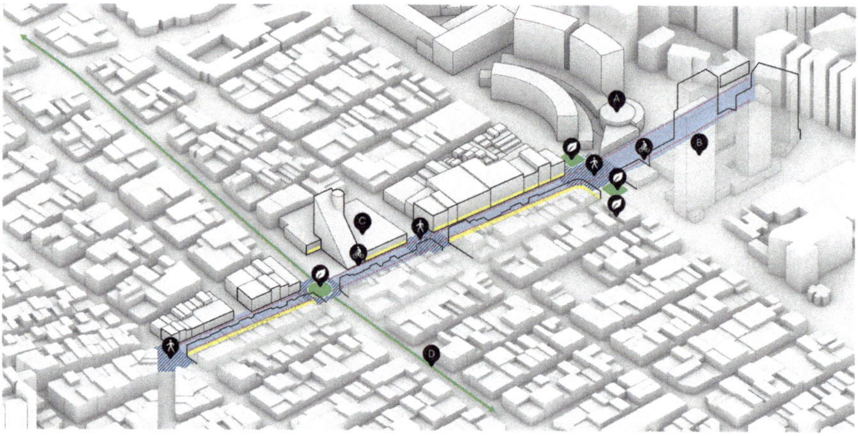

Fig. 6.16 Section 3 strategy

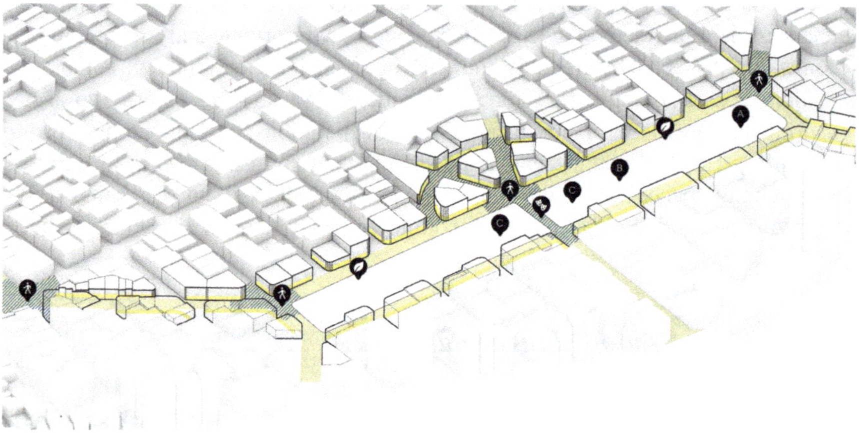

Fig. 6.17 Section 4 strategy

6.5.4 Fourth Section: Green Corridor

The final segment, the green corridor, spans Campos Elíseos, Arquímedes, and Andrés Bello streets, linking the project to Paseo de la Reforma. This transformation of Campos Elíseos, one of Polanco's most important streets, integrates it smoothly with the rest of the city through a green promenade that culminates in its junction with Paseo de la Reforma. This connection is not only symbolic but also practical, with a reorganization of vehicle traffic, expanded pedestrian spaces, and a continuation of the cycle path into the city's network (Fig. 6.17).

Arquimedes Street is redesigned as a shared plaza, enhancing pedestrian safety through restructured traffic flow and appropriate landscaping, providing urban furniture for user comfort. On Campos Eliseos, safe crossings are created at every corner, with sidewalks extended to reduce pedestrian exposure. These expanded corners generate rest areas, framed by low vegetation and Acacia Azul trees, highlighting their presence and contributing to the area's overall green infrastructure.

6.5.5 *Resilience by Design: Addressing Dry Urbanism Through Integrated Urban Strategies (Fig. 6.18)*

1. **Sidewalks:** Widening sidewalks improves the quality of pedestrian environments and facilitates movement. By providing ample space for pedestrians, this strategy enhances safety, encourages walking, and reduces congestion. It also supports the integration of urban design elements such as seating and landscaping, crucial for adapting to changing climate conditions.
2. **Cycling infrastructure:** The development of dedicated cycling lanes promotes sustainable mobility by offering a safe and efficient alternative to motorized transportation. This strategy not only encourages bicycle use but also helps mitigate traffic congestion and reduce carbon emissions, aligning with sustainable urbanism principles in response to increased aridity.
3. **Infrastructure:** The implementation of durable, aesthetically integrated subterranean facilities, such as drainage systems, enhances the urban environment. By keeping essential infrastructure concealed, this approach preserves visual appeal and ensures functionality, thus contributing to a resilient urban infrastructure capable of withstanding drier conditions.
4. **Infiltration planters:** The use of infiltration planters captures and filters stormwater, aiding in the replenishment of aquifers. This strategy addresses water scarcity by improving water management and reducing runoff, thus supporting

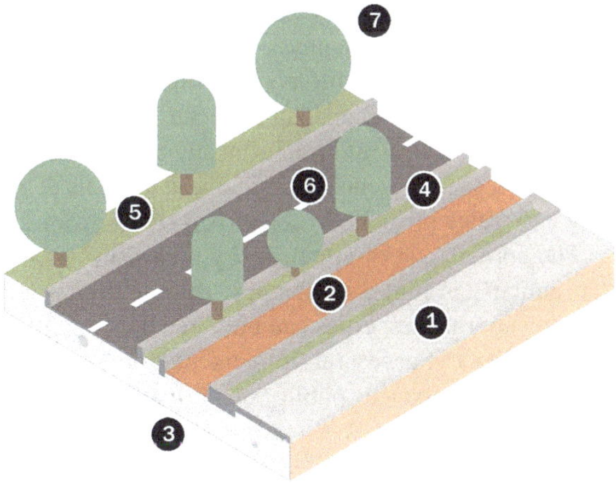

Fig. 6.18 Resilience by design

sustainable urban water cycles and enhancing resilience in the face of dry urbanism.

5. **Vegetation:** Strategic incorporation of vegetation organizes urban spaces and mitigates the urban heat island effect. By providing shade and cooling the environment, vegetation improves thermal comfort, air quality, and aesthetic appeal, making urban areas more adaptable to the impacts of climate change.
6. **Pavement materials:** The use of high-durability, low-maintenance pavement materials ensure long-term performance and reduces maintenance requirements. These materials contribute to the resilience of urban infrastructure by providing reliable surfaces that can withstand the increased wear and tear associated with drier climates.
7. **Trees:** The integration of trees enhances urban legibility and unifies cityscapes while promoting biodiversity. Trees offer essential ecological benefits, including habitat provision and air quality improvement, and contribute to creating a pleasant and sustainable urban environment, vital for adapting to dry urbanism.

Paseo Polanco's multi-faceted approach offers a comprehensive solution aligned with dry urbanism principles, emphasizing the creation of sustainable, resilient urban environments that adapt to changing climatic conditions while fostering a vibrant cultural and commercial atmosphere. The integration of green infrastructure, pedestrian-friendly spaces, and alternative mobility solutions ensures that Paseo Polanco not only responds to the needs of today but also anticipates the challenges of the future, making it a model for urban development in dry and diverse climates alike.

6.6 Conclusion

In the face of escalating climate challenges, particularly in arid environments, the urgent need for adaptive urban strategies is undeniable. Climate change exacerbates droughts and extreme temperatures, jeopardizing urban infrastructure and quality of life. Dry urbanism, encompassing effective water management, sustainable design, and equitable planning, offers a comprehensive response.

The Master Plan Ciudad Juárez and Paseo Polanco exemplify dry urbanism in action. By strategically configuring city blocks, prioritizing green spaces, and focusing on pedestrian-oriented design, these projects demonstrate how thoughtful planning can create resilient, livable urban environments.

The cornerstone of effective dry urbanism lies in adaptive, site-specific design that anticipates future challenges. By combining local knowledge with forward-thinking approaches, cities can build resilience, improve livability, and mitigate climate impacts. Successful implementation of dry urbanism strategies has the potential to enhance public health, stimulate economic growth, and foster social equity.

To fully realize the potential of dry urbanism, concerted efforts are needed. Investing in research, development, and public awareness is crucial. While challenges exist, the long-term benefits in terms of environmental sustainability, public health, and economic prosperity make dry urbanism a compelling strategy for cities worldwide.

References

CNN (2024, February 25) Retrieved from CNN: https://edition.cnn.com/2024/02/25/climate/mexico-city-water-crisis-climate-intl/index.html

El Economista (2024, February 21). Retrieved from https://www.eleconomista.com.mx/politica/Suministro-diario-de-agua-en-CDMX-se-ha-reducido-50-en-25-anos-20240221-0026.html

FAO (2022) The state of food and agriculture 2022. Leveraging automation in agriculture for transforming agrifood systems. United Nations, Rome

Farshid Aram EH (2019) Urban green space cooling effect in cities. Heliyon 5:e01339

IPCC (2021) Climate change 2021: the physical science basis. Intergovernmental Panel on Climate Change

Taleb HM (2014) Using passive cooling strategies to improve thermal performance and reduce energy consumption of residential buildings in U.A.E. Buildings. Front Archit Res 3(2):154–165

UN (2018) World urbanization prospects: the 2018 revision. Retrieved from https://population.un.org/wup/. United Nations

UN (2022) World cities report, chapter 10, building resilience for sustainable urban futures. United Nations

World Bank (2023) Enablers of inclusive cities: enhancing access to services and opportunities. World Bank

WRI (2019) Aqueduct 3.0: updated decision-relevant global water risk indicators. Retrieved from https://www.wri.org/publication/aqueduct-30. World Resources Institute

Zhang F, Borja-Vega C (2024) Water for shared prosperity. World Bank, Washington, DC.

Chapter 7
Urban Arid Green: A Nature-Based Solutions Proposal for Ecocity Development in Arid Regions. Case Study Tamansourt, Morocco

Rosa de Wolf, Nico Tillie, Rob Roggema, and Kristel Aalbers

Abstract Drylands, hyper arid to subhumid areas where rainfall is limited, are expected to expand due to climate change. Natural resources, such as water and food, are scarcer in these areas. Population growth and urbanization are putting even more pressure on communities living there, as well as on the urban fabric and ecosystems. How can nature based ecocities be created in these environments?

A pattern language, which is able to translate practical knowledge into substantiated spatial configurations that work in arid areas is missing in current theory and practice. The ´Urban Arid Green research by design project´ addresses a sustainable population growth and urbanization in arid regions via the case study in Tamansourt. Tamansourt is one of the 19 Moroccan new towns developed under the national Villes Nouvelles (New Towns) strategy. The city is still under construction, as the spatial analysis, site visit, and conducted interviews have revealed. The city has not reached its target population nor its desired level of urban activity yet. However, fundamental issues already manifest themselves.

The vision 'Regreen to Rewild' aims to counteract the pressures mentioned before, by developing a transformative framework towards an ecocity, taking the natural system as the basis (nature-driven urbanism). Tamansourt Ecocity gives purpose to the local community and a new identity to the city.

The transition described in this vision requires a systemic change, including environmental, policy and behavioral change. To support this shift, a common language amongst all future stakeholders has been created: the Urban Arid Green pattern language. This spatial language includes four pathways, of which each includes a group of development guidelines. One of the four pathways captures the original

R. de Wolf (✉) · N. Tillie · K. Aalbers
Faculty of Architecture and the Built Environment, Delft University of Technology, Delft, the Netherlands
e-mail: R.deWolf@tudelft.nl

R. Roggema
Escuela de Arquitectura, Artes y Diseño, Tecnológico de Monterrey, Monterrey, Nuevo León, Mexico

R. Roggema (ed.), *Dry Urbanism*, Contemporary Urban Design Thinking,
https://doi.org/10.1007/978-3-031-77347-1_7

design guidelines for Moroccan new towns, as formulated by Al Omrane in 2010. The other guidelines focus on the biosphere, the urban fabric, and immaterial values.

The Urban Arid Green project shows how these generic patterns can be translated into a site-specific design which stakeholders can use in the city's transition. This sets a precedent for other arid cities that aim to sustainably develop while under the pressure of scarce resources, climate change and population growth. Every landscape needs to adapt its own site-specific pathway based on the generic language, allowing unique dialects for different landscapes.

Keywords Ecocity · (un)sustainable growth · Aridity · Natural resources · New towns · Spatial nature-based solutions · Pattern language

7.1 Introduction

The global population is growing, projected to increase from 7.7 billion in 2019 to 10.9 billion by 2100 (Maja and Ayano 2021). Cities today face significant challenges due to this growth, with the largest increase expected in low-income countries where societies are more dependent on natural resources. This population surge, coupled with urbanization and economic development, escalates the demand for food, water, and energy. Drylands (Fig. 7.1) range from dry subhumid to hyper-arid areas. Currently, over two billion people live in drylands, which are expected to expand due to climate change (Rajaud and Noblet-Ducoudré 2017).

Many economies in arid regions rely heavily on agriculture, but climate variability threatens crops with excessive, insufficient, or untimely rainfall (Keeton and Provoost 2019, p. 34). The increasing demand for irrigation increases water

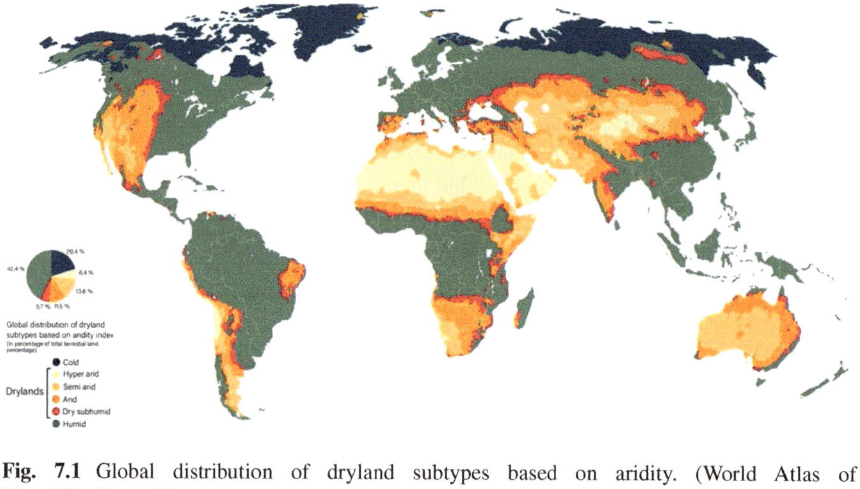

Fig. 7.1 Global distribution of dryland subtypes based on aridity. (World Atlas of Desertification n.d.)

insecurity (Hill and Pimentel 2022). These arid areas already suffer from limited access to natural resources, particularly water. Consequently, the overall wealth and health of societies in drylands are lower than those in more humid areas (Rajaud and Noblet-Ducoudré 2017). The African continent, especially its northern region, is predominantly covered by drylands.

Urbanization and population growth worldwide create spatial challenges in cities, which can be addressed by densifying or expanding existing cities or building new urban settlements. These so-called New Towns are found in many countries, particularly in rapidly urbanizing Africa, which experiences significant population and urban growth. However, many new towns in Africa, are vulnerable to climate variability and climate change (Keeton 2020). Moroccan new towns face serious desertification risks (Fig. 7.2).

Morocco's rapid population, urban, and economic growth has drastically changed its urban landscape, leading to a housing shortage. To manage uncontrolled urbanization and support economic growth, Morocco implemented the Villes Nouvelles strategy, founding nineteen new towns (Côté-Roy and Moser 2022). The geographical distribution of these towns is shown in Fig. 7.3.

Tamansourt, Tamesna, Lakhyayta, and Chrafate are the first-generation satellite new towns, established between 2004 and 2009 (Côté-Roy and Moser 2022, p. 34). Building a new urban structure from scratch is a long-term process, meaning these cities do not immediately meet all demands. After 20 years, it's time to evaluate

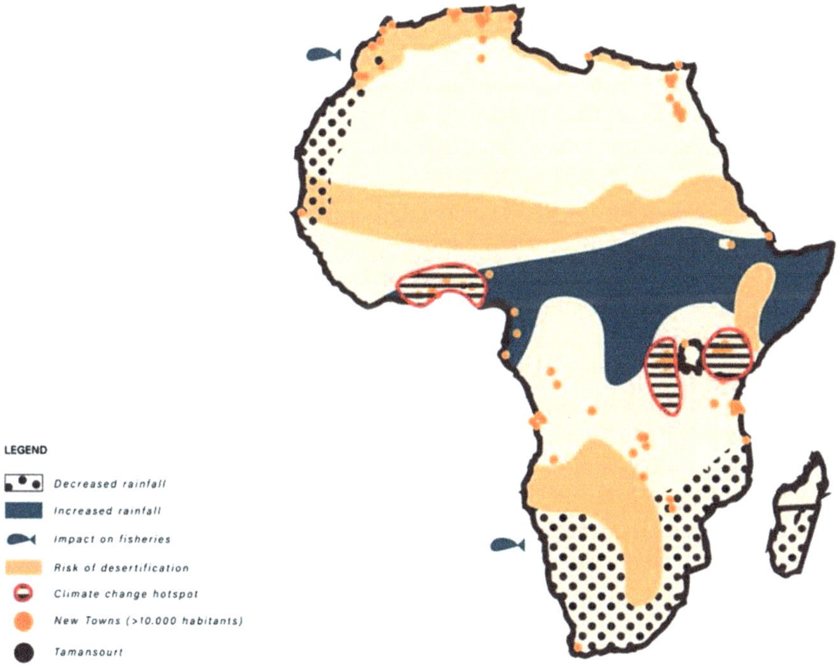

Fig. 7.2 Climate change threats Africa. (Keeton 2020, p. 44) adjusted by author

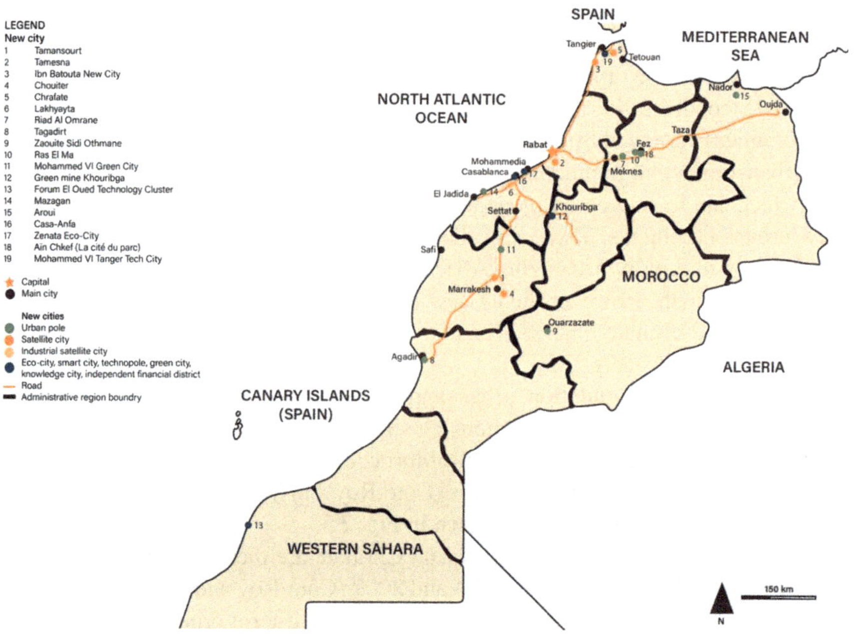

Fig. 7.3 Geographical distribution of new cities in Morocco. (Côté-Roy and Moser 2022, p. 30) adjusted by author

these first-generation new towns. The first-generation new towns are designed as satellite cities, and there is now an aim to transform them, along with three other towns, into ecocities. This transformation project is led by Al Omrane, the state company that developed these cities (Ecocity Morocco n.d.).

Tamansourt, established in 2004 to prevent overpopulation and congestion in Marrakech, stands out for its large area and target population of 450,000 inhabitants. Tamansourt's landscape is the most barren (Fig. 7.4). Located in the Tensift Basin, the area faces limited groundwater and surface water availability, with occasional stark seasonal contrasts of heavy rainfall and extreme droughts (Fniguire et al. 2014; Abdelaziz et al. 2022).

People moved to Tamansourt for better access to urban amenities and job opportunities. However, despite a focus on social and affordable housing, the city lacks adequate job opportunities and urban amenities such as public transport. The goal was to create a low-density, green city with a harmonious mix of residents, but Tamansourt, called a dormitory city, has not yet achieved these goals. Tamansourt's target density of 23100 inhabitants/km^2 (Côté-Roy and Moser, 2022, p. 30) exceeds the by UN-Habitat considered ideal density of 15000 inhabitants/km^2 (Keeton and Provoost, 2019, pp. 92, 111). Tamansourt is located 10 km northwest of Marrakech, separated by the Tensift River (Fig. 7.5). The Marrakech-Safi highway allows a 15-min commute by car. Many residents commute to Marrakech for daily activities.

Over time, Tamansourt's land degraded. Aerial views from 1985 to 2021 show the disappearance of green wadi structures (Fig. 7.6), now functioning only as

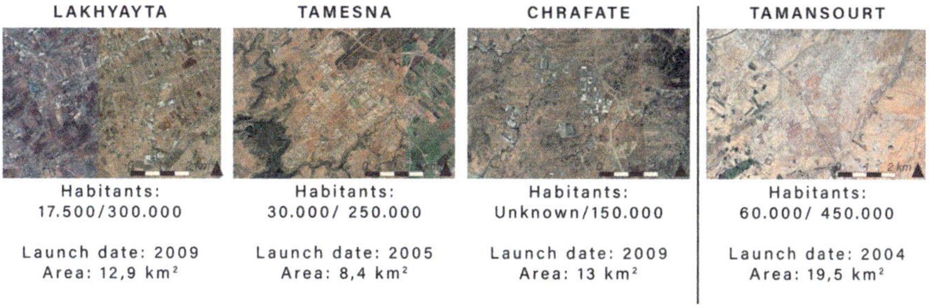

Fig. 7.4 Comparison between first generation of New Towns. Images retrieved from (Google Earth Pro 2022), facts from (Côté-Roy and Moser 2022, p. 30)

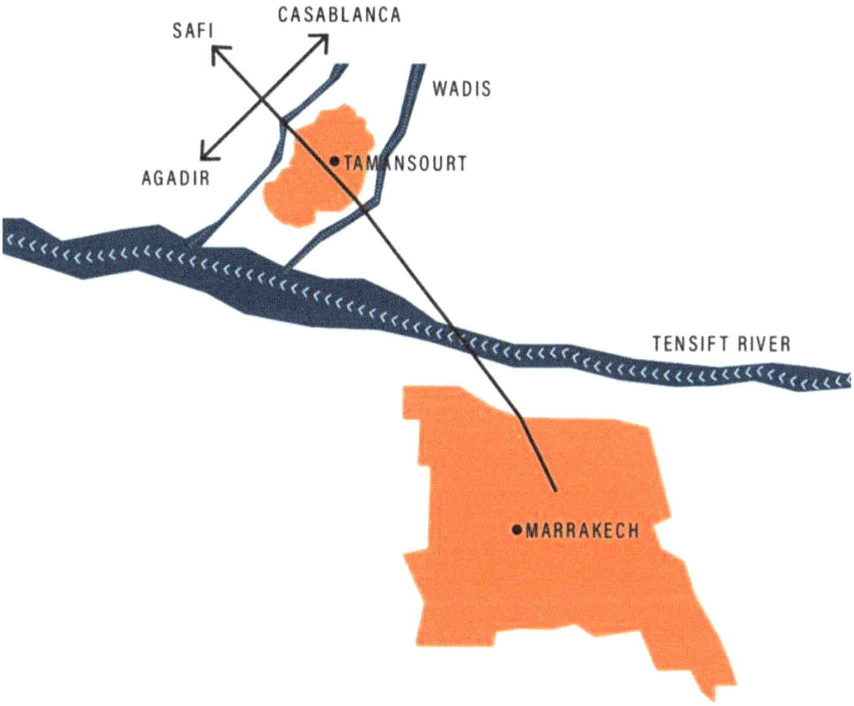

Fig. 7.5 Schematic situation of Tamansourt

erosion gullies. The landscape and ecosystem degradation coincides with the city's focus on built environment development.

As Tamansourt densifies further, unsustainable development could exacerbate degradation. Transforming Tamansourt into an ecocity, as proposed by Al Omrane, requires spatial redevelopment. This chapter examines how the ecocity principles can be translated into a spatial design that fosters sustainable interactions between human and natural systems. Tamansourt had a notable population of young children

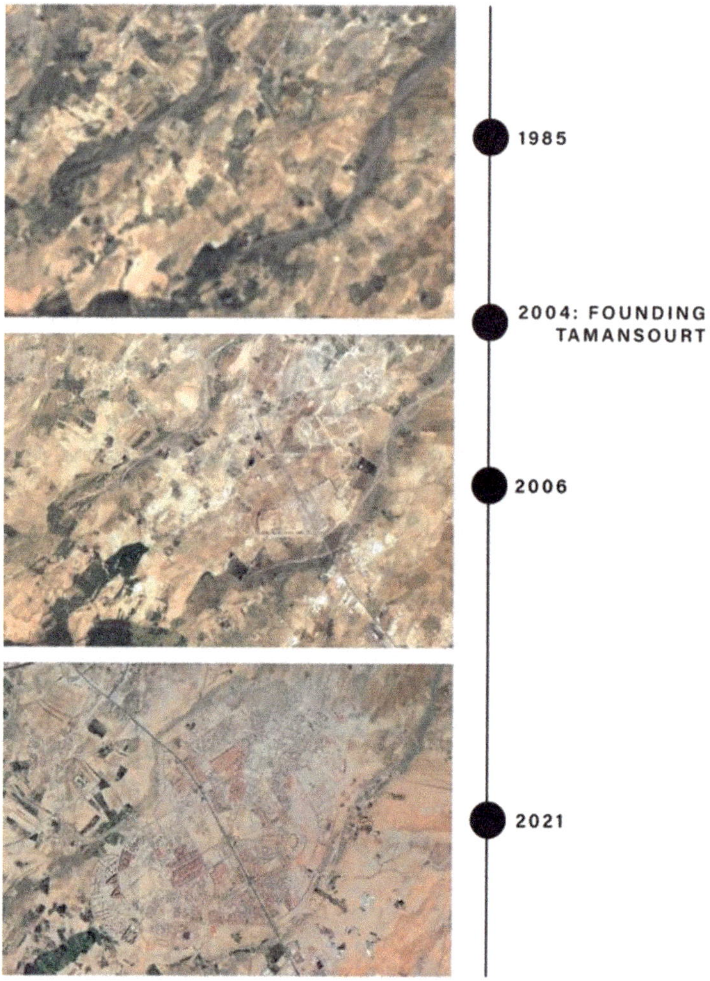

Fig. 7.6 Timeline development site of Tamansourt. (Images retrieved from Google Earth Pro 2022)

in 2015, indicating large families. The median age was 24.6 years, with a nearly equal gender ratio (City-facts n.d.). Tamansourt aims to accommodate demographic growth, control urban sprawl, and stimulate economic revival. The town is designed for low-density living with a variety of housing types to meet diverse needs (Abdelaziz et al. 2022).

As in most Islamic towns, clear sightlines to the mosques result in predominantly low-rise buildings in Tamansourt. This aimed target population leaves little open public space, which currently is mainly occupied by wide asphalt strips for traffic. Unlike traditional Moroccan towns, Tamansourt lacks historical layering. The city's current construction status shows finished areas and ongoing development, with limited activities and job opportunities, mostly in the construction sector. Future plans include a university campus, an industrial site, and more social housing, aiming for a diverse social and economic mix. Tamansourt is located in the Tensift Basin, prone to soil erosion and land degradation. The arid and rocky soil, combined with a lack of vegetation, exacerbates water runoff and soil loss, leading to flooding and droughts. To address these issues, urban design must focus on land conservation and soil restoration. Morocco faces a severe water shortage, expected to worsen by 2030. The Green Morocco Plan has enhanced agricultural output but increased social inequalities and dependency on unpredictable rainfall. Generation Green 2020–2030 aims to modernize traditional farming and improve resilience to climate change. Sustainable agriculture is essential for soil and water conservation.

Vegetation significantly cools the urban environment and improves residents' mental and physical health. The IUCN recommends integrating more native species to enhance ecosystem services like these (Ecocity Builders 2022). Public green spaces are crucial to Tamansourt's vegetation coverage, as there are limited private gardens with this predominantly social housing stock. Tamansourt's design must prioritize sustainable development by incorporating more vegetation and focusing on land and water conservation. This approach aligns with Al Omrane's vision and the Ecocity Builders' framework and standards (AL Omrane Marrakech 2008; Ecocity Builders n.d.), promoting a comfortable urban climate and resilient infrastructure.

7.2 Objective

As a long-term objective, Tamansourt will have been transformed into an ecocity by 2040, and shifted from environmental degradation to ecological restoration. The city leverages nature to foster sustainable development, with active regreening allowing the landscape to regenerate and rewild (Fig. 7.7). This transformation creates a balanced urban landscape where nature, agriculture, and human activities coexist, contributing to a resilient, healthy biosphere. New opportunities arise through plant nurseries, ecotourism, and agrotourism, fostering economic and social independence. Training and employment programs ensure that residents actively participate in this transition, promoting a sustainable economy.

The ecocity concept extends beyond Tamansourt, inspiring regional rewilding and reversing desertification. The transition focuses on closing essential loops, particularly in water management and local food production. A circular water system is established, using treated wastewater to restore soil quality and support vegetation. Tamansourt becomes a testing ground for innovative agricultural practices, functioning as a botanical garden with educational tourism. Central and local hubs produce seedlings for regreening efforts, involving residents in the greening process. The city plans to regreen its wadis, creating recreational spaces and additional water sources.

The ecocity transition enhances Tamansourt's economy, diversifying from construction to sectors like tourism, agroforestry, and education. Key locations and

Fig. 7.7 Regreen to Rewild: Tamansourt Ecocity 2040

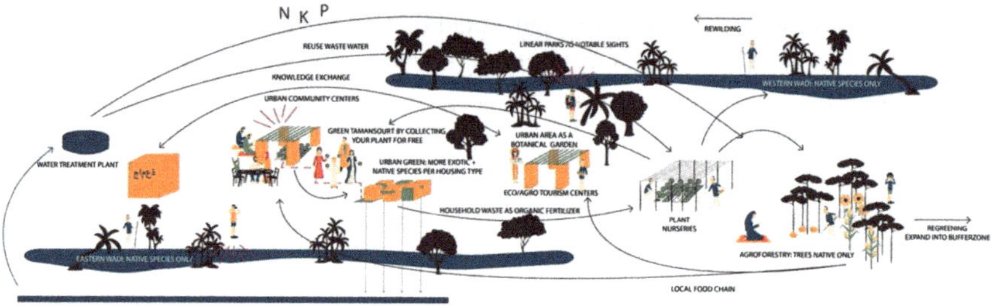

Fig. 7.8 Close the Loops

hubs stimulate systemic change, spreading to the region between Tamansourt and Marrakech. For a sustainable future, Tamansourt must optimize resource use and rethink its interaction with nature, focusing on water retention through vegetation (Fig. 7.8). The transformation involves a circular water system, top-down support from relevant ministries, and active local participation. Empowering private actors and residents ensures the city's self-sufficiency and financial sustainability.

Community involvement is crucial, with education and awareness programs to promote environmental stewardship. Engaging universities, developers, and local farmers in the transition fosters a trust-based partnership for a sustainable Tamansourt ecocity.

It is crucial to train local farmers in agroforestry techniques and to develop a tourism program that integrates eco- and agrotourism activities. In addition to training, it is essential to create employment opportunities in these sectors. Both short-term and long-term successes are necessary for sustainable urban development. By 2040, the local community envisions living in an ecologically healthy city, enhancing livability and comfort. In the short term, immediate rewards such as income generation and increased green spaces are also pivotal. Initially, a combination of

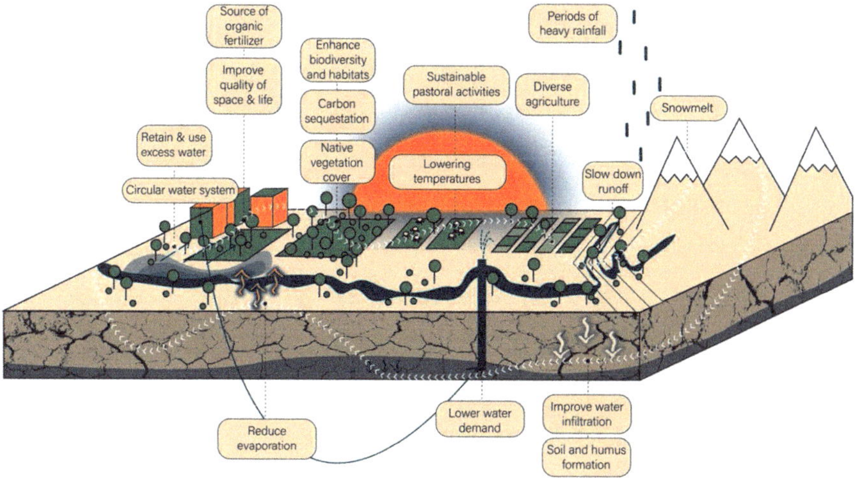

Fig. 7.9 Environmental change

incentives and regulations may be required to ensure that human activities do not disrupt the natural environment and allow for ecosystem recovery (Fig. 7.9).

7.3 Approach

This chapter searches for an approach wherein urban design can contribute to solving the hereabove named problems by sustainably addressing population growth and urbanization in arid areas.

Sustainable urbanization can take place when the spatial condition for agglomeration, accessibility and quality of life are considered (Ram and Gerretsen 2014) (Fig. 7.10). Taking these spatial conditions as a driver in the design of the urban landscape, economic developments can be further stimulated while strengthening the competitive position of the site.

In 2015, the 2030 Agenda for Sustainable Development was launched, providing a globally shared master plan for sustainable development specified in seventeen Sustainable Development Goals (SDGs). These goals serve as a global call to action for building peace and prosperity (UN 2015). The Stockholm Resilience Centre (2016) integrated the SDGs into a ´wedding cake model´ in which the biosphere functions as the foundation for society and the economy and is therefore essential for achieving global sustainability (Stockholm Resilience Centre 2016). When designing for sustainable urban development, this biosphere should be recognized as the basis. Landscape-based urbanism is "a multi-scale planning and design approach for developing resiliency and adaptive capacity by creating flexible, strong socially and ecologically inclusive landscape structures, connecting long-term

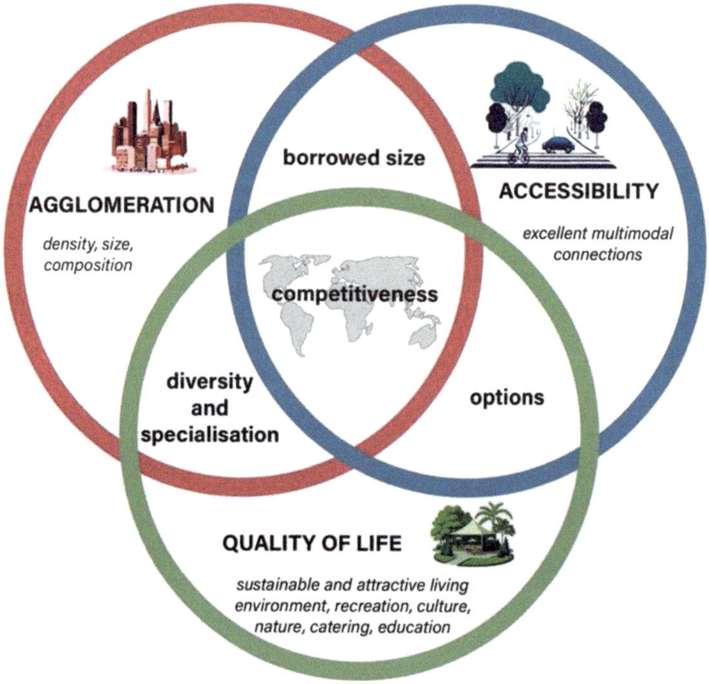

Fig. 7.10 Coherence between agglomeration power, accessibility, and quality of life, (Ram and Gerretsen 2014). Translated and adjusted by author

perspectives with short-term interventions using research through design as wayfinding in a transdisciplinary process." (Nijhuis 2022, p. 251; Roggema 2020), The concept of the ecocity aligns with this landscape-based urbanism approach. Ecocity Builders (2010) define an ecocity as "a human settlement modelled on the self-sustaining resilient structure and function of natural ecosystems. An ecocity provides healthy abundance to its inhabitants without consuming more (renewable) resources than it produces, without producing more waste than it can assimilate, and without being toxic to itself or neighboring ecosystems. Its inhabitants' ecological impact reflects planetary supportive lifestyles; its social order reflects fundamental principles of fairness, justice, and reasonable equity."

These concepts form the backbone of the conceptual framework of the Urban Arid Green project (Fig. 7.11), which sustainably addresses population growth and urbanization in arid areas as a systemic solution to global sustainability. It presents urban development as a key driver affecting all linked systems within the landscape, including socio-cultural, biological, and physical layers. By building with nature, the Urban Arid Green project interlinks the productive landscape, urban area, and waterscape with these systems, aiming to create great, fair, and resilient places that form an ecocity. When ecocities work together as a network, they contribute to

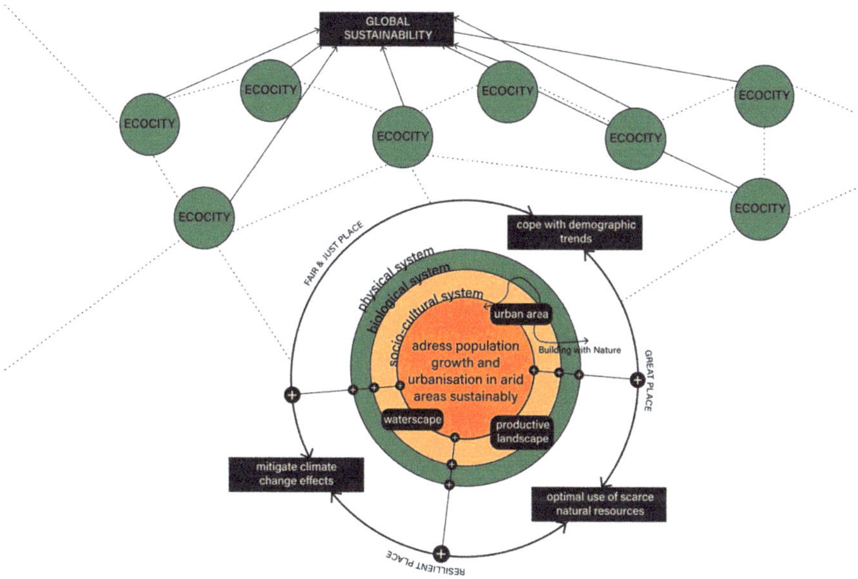

Fig. 7.11 Conceptual framework Urban Arid Green

global sustainability by coping fairly with demographic trends, mitigating climate change, and optimizing the use of scarce natural resources.

The Urban Arid Green project aims to demonstrate how the demand for housing and natural resources can complete rather than compete. Urban Arid Green seeks to integrate human, natural, and agricultural activities in urban areas, focusing on arid regions to support growing societies facing resource disparities. Tamansourt, like many arid new towns, faces significant challenges. The project envisions an ecologically healthy and vibrant Tamansourt by reversing the destruction of the natural system. Tamansourt will be used as a case study to formulate and test guidelines for designing with limited natural resources, serving as an example for other new towns in arid regions when generalizing the process for applicability to these other locations. This holistic approach aims to counteract the growth of arid areas and desertification while mitigating climate change. Based on the analysis and dialogues on-site, a vision for Tamansourt, "Regreen to Rewild, Tamansourt Ecocity 2040," has been designed, drawing on Al Omrane's original vision and the Ecocity Builders' principles. The planned built volumes are retained, with systemic environmental, policy, and behavioral changes needed for this transition defined. Given that many cities face similar pressures from climate change, population growth, urbanization, and scarce resources, a strategy useful for other cities is developed. The Urban Arid Green pattern language is formulated, incorporating guidelines by Al Omrane (2010) and additional pathways: Circularity & Ecology, Supportive Lifestyles, and Facilitating Fabric. These socio-spatial guidelines focus on supporting the transition into an ecocity. Validation of this pattern language via a design exploration in

Tamansourt sets an example for other urban arid landscapes that aim to support population growth and urbanization sustainably.

7.4 Application and Results

The concept of a pattern language, as described by Alexander in "A Pattern Language: Towns, Buildings, Construction" (1977), provides a methodological framework for sustainable urban design. It consists of a series of interconnected patterns applicable to various scales, ensuring cohesive development across buildings, neighborhoods, and cities. This chapter explores sustainable approaches to managing population growth and urbanization in arid regions, focusing on Tamansourt as a case study. Effective implementation of the proposed Urban Arid Green language necessitates early engagement of all stakeholders to ensure comprehensive understanding and support (Fig. 7.12).

To transition Tamansourt into an ecocity, socio-spatial design guidelines emphasizing biosphere sustainability are crucial (Fig. 7.13). These guidelines integrate insights from Al Omrane's design principles with new patterns focused on circularity, ecology, supportive lifestyles, and urban fabric facilitation. The transformation of Tamansourt involves reorienting its urban fabric from being Marrakech-centric only to a more self-contained, community-oriented structure as well. This shift includes improving local infrastructure and creating vibrant communal spaces like seasonal parks and central hubs for agriculture, education, and tourism. Addressing infrastructural needs, such as redesigning major roads like the Marrakech-Safi highway, promotes safer, more accessible urban mobility. Enhancing these corridors as integrated hubs encourages active community engagement and economic vitality. In conclusion, the Urban Arid Green initiative aims to redefine urban development in arid areas, fostering sustainable growth and enhancing community resilience. By

Fig. 7.12 All stakeholders speak the same language, they understand each other and work together on the transition towards an ecocity

integrating local knowledge with innovative design strategies, Tamansourt can evolve into a model ecocity, benefiting both current and future generations.

The analysis shows that Tamansourt is perceived as a dormitory city, which is lacking sufficient urban amenities and job opportunities. The urban fabric of Tamansourt supports this, it is Marrakech-oriented. For Tamansourt to lose this image, its structure must shift from being Marrakech-oriented only, to becoming Tamansourt-oriented as well. The vision for Tamansourt as designed by Al Omrane Marrakech (2008) is altered to the Tamansourt-oriented plan. Key aspects of this plan are the mobility shift, local hubs, central hub, seasonal parks, wastewater treatment, and appropriation.

The mobility shift improves accessibility (Fig. 7.12). Tamansourt is designed to be dense, therefore the urban fabric between buildings and their interaction is important to the spatial quality. As the open public space primarily comprises infrastructure, a shift in mobility is necessary. Currently, the Marrakech-Safi highway runs through the core of Tamansourt. Pedestrian bridges allow slow traffic to pass over this fast traffic. The highway splits into 14-meter-wide asphalt avenues, dividing the city and creating both physical and non-physical barriers. By prioritizing the outer road for entering or crossing Tamansourt, slow traffic within the city is given precedence and enhanced comfort. This approach reduces the infrastructure's barrier effect within the urban layout. Prioritizing slow traffic and shared mobility will foster safer, healthier, and more social mobility. As a result of this shift, the city's grand avenues can evolve from mere traffic arteries to vibrant urban axes of activity, connecting various centers. The outer road will link the rewilded natural landscape or agroforestry areas to industry, businesses, the university campus, and housing with retail. The design principle for the outer road emphasizes safety. Here, slow

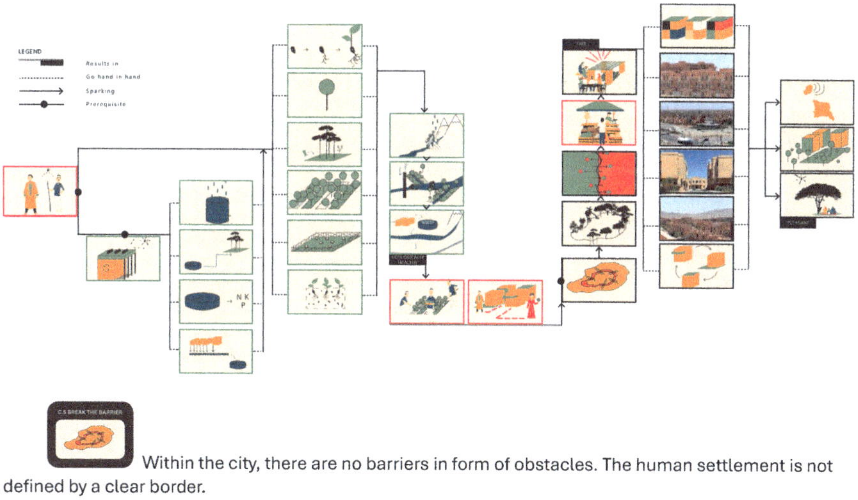

Fig. 7.13 Urban Arid Green Pattern Language

Development must ensure an improvement of the quality of space.

Spaces for common interest around the city are available to everyone.

Safe & comfortable spaces allow people to come together in public space.

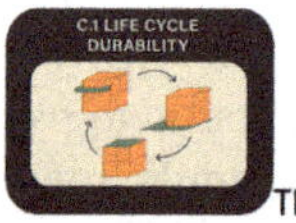

The urban form of the city allows inhabitants to grow through different stages of life.

Identity must be created. Although a city must look as one whole, the city allows individuality.

Entities are created throughout the city.

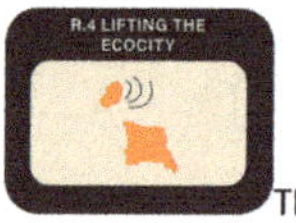

The idea of the Ecocity is lifted to the regional scale.

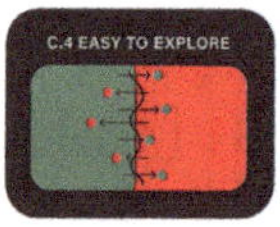

The urban form is super clear. It is easy and safe to explore areas one does not necessarily has to go to.

Businesses are functioning via a low-profile business operation. Over-modern methods, equipment and technology is omitted.

Fig. 7.13 (continued)

The financial system is fair.

The population acquires know-how by education and training. This must be open to all, disregard one's status.

In the urban area, vegetation is included in every private plot. On the plot, this can be done in various ways via greening gardens, façades, balconies and roofs.

Wastewater is treated. This can be done via centralized and decentralized methods, on different scales.

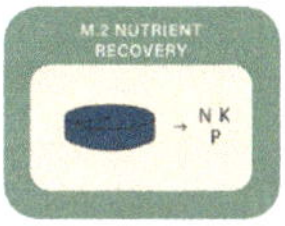

Nutrients are recovered from wastewater. This can be done via centralized and decentralized methods on different scales.

On agricultural land, perennials are mixed with annual crops.

Create healthy soil ecosystems with techniques as cover cropping, residue mulching, conservation agriculture, phase rotations and creating continues vegetation cover.

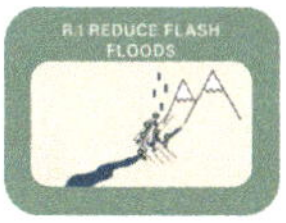

Flash floods must be reduces by infiltrating water as soon as possible. Downstream, room for the river is made to lower peaks in discharge.

Grow native species to green the environment

Fig. 7.13 (continued)

Livestock/wildlife is not repeatedly grazing of vegetation at such pace that the vegetation does not have the time to regrow.

Aim for closed cycles on every level, that have as little negative impact on the environment as possible.

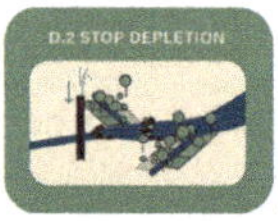

Stop depletion of the aquifer by lowering the water intake from the aquifer on agricultural land.

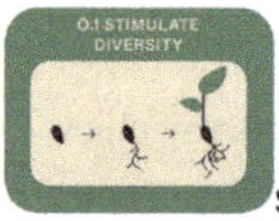

Support genetic diversity amongst vegetation, to stimulate its resilience.

On green spaces, only organic fertilizer is used, made out of household waste or nutrient recovery.

Excess water is captured and stored for later purposes.

Water reuse is safe, following possible restrictions as for instance covered irrigation only.

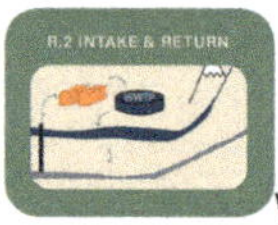

Water is not only taken from the natural system. Clean water is also returned to nature, replenishing the aquifer.

Urban, landscape and architectural design refer to local architecture in a modern way. High maintenance is avoided.

Fig. 7.13 (continued)

Buildings interact without seeking symmetry. There is a diversity in architecture between blocks, but unity within blocks.

The hierarchy in routes is clear and expandable. Different functions in different districts. Non-residential buildings are more expressive.

The new towns are linked to other cities. They adapt to the relief and are integrated to the geographical and historical site. Their territory allows further development.

Fig. 7.13 (continued)

and fast traffic are separated, adequate lighting is provided, pedestrians are prioritized, and bus stops are strategically located. Since the outer road is not yet built, its layout can be reconsidered and optimized (Fig. 7.14).

By qualitative densification of the fabric, the agglomeration power strengthens. The urban area is activated with spaces for common interest in the three, by Al Omrane appointed, centers. This allows specification and diversification of functions. The central hub is located on the northeast side of Tamansourt, outside the planned outer road. Here, food production, education, research, and touristic activities synergize.

The quality of life is further improved via the greened wadis enclosing the ecocity, that translate the old Moroccan medina walls into those of a new ecocity. Creating two linear parks for all inhabitants to enjoy. To activate this new ecocity framework, spaces for social gathering are created. These are located within the seasonal parks and central hub on the activity strip, on the continuation of the main axis in the city.

The local hubs provide essential services to enhance community life. At the shop, locally grown food from the central hub is available for purchase, ensuring fresh produce for residents. Nurseries cultivate urban greenery, encouraging locals to participate in city greening efforts. Cafes and restaurants utilize local ingredients, promoting economic opportunities and fostering social interactions among residents. These centers serve as focal points for community engagement and empowerment. Addressing deficiencies identified in the analysis, each hub in Tamansourt is tailored to meet specific community needs. This customization ensures that amenities align closely with local demands and complement existing functions within the hub's vicinity. Variations between hubs reflect unique characteristics and priorities. The design of local hubs is influenced by the visionary model at the Al Omrane Tamansourt office with adjustments made to accommodate existing built structures where necessary.

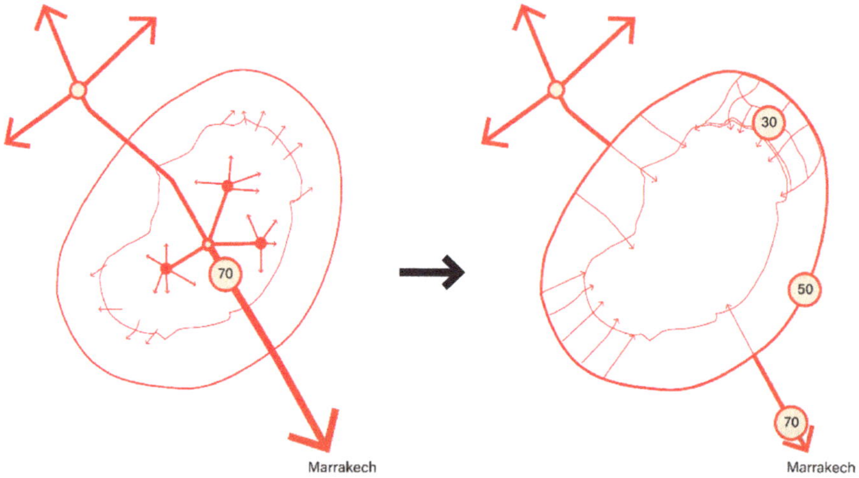

Fig. 7.14 Rethinking the mobility of Tamansourt

The analysis highlighted existing green avenues within Tamansourt's infrastructure, crucial for maintaining pleasant boulevards. The vision for Tamansourt in 2040 involves extensive community involvement in urban greening efforts. Based on Al Omrane's zoning plan, residential areas with shared courtyards aim to maximize ground-level green spaces. Buildings without outdoor areas prioritize rooftop and vertical gardens. This diverse approach to urban greening enhances mental well-being, urban climate, and overall quality of life. The selection of plant species for Tamansourt's built environment includes native drought-resistant species alongside a variety ornamentals within the inner-city. The southern hub features public football and tennis courts, enhancing local recreational opportunities. These facilities bolster the area's sports scene and complement nearby amenities like parks and a mosque. Linked to a new residential neighborhood via the outer road, the central hub prioritizes pedestrian and public transport access. This hub serves as a gateway to the activity strip, promoting agroforestry and tourism. The hub's elevated position compared to the city, higher on the foot of the Jebilet, offers panoramic views of seasonal parks and forests. It will support educational and research activities.

The central hub pioneers perennial ecosystemic agriculture, transforming desertified areas into productive ecosystems. A diverse mix of native, drought-resistant trees and crops are cultivated, supported by a robust selection process. Governmental policies and local initiatives aim to sustain this agricultural transformation, ensuring food security and economic resilience. Tamansourt strives for a self-sustaining local food chain, promoting crop rotations and sustainable harvesting practices. Calculations must indicate the necessary land area for agroforestry to meet local dietary needs, emphasizing the community's commitment to food security and environmental stewardship. By focusing on community-based initiatives and sustainable urban planning, Tamansourt aims to achieve a harmonious balance between human development and ecological resilience.

Fig. 7.15 Agroforest formation

7.4.1 Formation of Agroforestry

The initial stages of creating the Agroforest in Tamansourt begins with tree planting and trenching to transform the barren landscape into a thriving food forest (Fig. 7.15). These trees provide shade, initiating microclimate adjustments. Subsequently, organic waste and recovered nutrients facilitate humus and soil formation, closing nutrient loops. Pruning and leaving branches further accelerate this process. Once conditions permit, crops are introduced between trees, marking the onset of agroforestry. As the forest stabilizes, harvesting of its products becomes feasible. Certain trees and crops, such as Quercus suber (Cork oak), require long-term planning, with harvests occurring after 25 years for these specific oaks, and every 9 years thereafter. Maintaining a detailed harvesting schedule ensures optimal workflow management.

The agroforestry site incorporates several design patterns, and shows the spatial result (Fig. 7.16).

7.4.2 Seasonal Parks

The seasonal parks play a pivotal role in preventing urban flooding in Tamansourt. By harnessing wadis more efficiently, these parks enhance city livability. Aggressive flash floods are managed through nature-based hydraulic solutions, essential for safeguarding erosion-prone gullies. Engaging communities residing along these gullies is crucial for Tamansourt's sustainable development. By transitioning from erosion gully to park, rainwater is allowed to seep into soil, replenishing reserves rather than causing erosion. The transition demonstrates key hydraulic measures of Tamansourt's waterbody. Slowing water flow and preventing overgrazing are primary strategies in restoring water bodies. Tamansourt's seasonal parks, exclusively feature native, drought-resistant flora that can withstand the fluctuating conditions of the wadi. This strategic selection minimizes maintenance costs while highlighting Morocco's indigenous biodiversity, boosting tourism and cultural significance.

The development of seasonal parks focus initially on livestock and water management, followed by natural regreening efforts and active afforestation (Fig. 7.17). Enhanced water infiltration and vegetation cover improve water safety and

Fig. 7.16 Impression Agroforestry Central Hub

Fig. 7.17 Phasing Seasonal Park

maintenance efficiency, creating local employment opportunities and fostering community engagement.

The potential outcome of the park shows the integrated design patterns (Fig. 7.18).

Seasonal parks encircle Tamansourt akin to historic Moroccan medina walls. Rammed-earth canopies strategically positioned throughout provide shade and social gathering spots, blending with the arid landscape. These canopies, drawing inspiration from Marrakech's El Badi Palace offer both shade and potential for solar energy generation, contributing to a sustainable urban environment.

Tamansourt's sustainable urban development integrates ecological restoration with community-centric amenities, fostering resilience against environmental challenges while promoting cultural heritage and local economic growth.

In 2024, the primary objective is clear: Tamansourt must transition into an ecocity. Achieving this goal requires robust top-down support to initiate data collection and foster open knowledge exchange. Collaboration among stakeholders within Tamansourt and neighboring new towns is essential. Local communities are energized and mobilized through this support and participate in finalizing the urban landscape design.

Locals can initiate decentralized wastewater treatment early on, a process that will evolve and strengthen with the introduction of tourist facilities. By 2030, as Tamansourt becomes more self-sufficient, plans for a wastewater treatment plant

Fig. 7.18 Impression of the Seasonal Park

(WWTP) can commence, followed by the reconstruction of sewage infrastructure upon completion. The WWTP's construction aligns with the finalization of sewage infrastructure to ensure effective water recovery. Simultaneously, nurseries and storage facilities can begin construction, and grazing control barriers established. These preparations facilitate seedling growth and kickstart active regreening efforts during rainy seasons, initiating the local food chain production. Once wadis and agroforestry sites are sufficiently regreened, the development of tourist facilities completes the activity strip. This phase introduces agro- and ecotourism to the area, prompting the formulation of a tourist program and generating new employment opportunities. Given the extended timeline for infrastructure design and reconstruction, phased implementation by district or project can start concurrently with the design phase. Local hubs can be constructed alongside infrastructure reconstruction. Educational and research centers can be established following the completion of the central hub's design. These centers, like a Faculty of Nature, Environment, and Agroforestry, utilize ongoing data collection efforts. Each phase's prerequisites ensure progression, allowing subsequent phases to commence upon achieving specific goals.

The project's adaptive pathway approach incorporates different sustainable scenarios toward achieving the ultimate goal. This flexible approach acknowledges uncertainties such as climate change impacts, socio-economic challenges, and external factors. It avoids linear processes and ensures continuous reflection and adaptation. Shared understanding and awareness among stakeholders are fostered. Workshops are pivotal in creating a socially safe environment for discussing actions and committing to common goals, despite any potential setbacks.

7.5 Discussion and Conclusion

Tamansourt's journey toward becoming an ecocity by 2040 underscores Morocco's strategic response to urbanization and population growth through new town expansion. The Urban Arid Green project exemplifies how urban landscapes can restore natural systems, integrating a green framework to foster sustainable interactions between humans and the environment.

Key to this transformation are systemic changes encompassing environmental, policy, and behavioral dimensions, addressing soil degradation and environmental challenges exacerbated by climate change. The project empowers local communities through collaborative efforts and bottom-up initiatives, ensuring the co-creation of a regenerative urban landscape.

The Urban Arid Green pattern language guides development by spatializing theoretical guidelines and aligning with the UN's Sustainable Development Goals, prioritizing sustainability at its core. As Tamansourt transitions into an ecocity, lessons from this project are applicable to similar landscapes, promoting communication and inspiring global sustainability efforts.

In managing scarce resources like water, food, and energy, urban design plays a pivotal role in creating circular systems that optimize resource use while fostering equitable social and economic structures. Local engagement is central to the ecocity's success, ensuring sustainable urban development that adapts to future uncertainties.

Ultimately, Tamansourt's journey represents a model for sustainable urban development in arid regions, emphasizing adaptability, resilience, and community-driven initiatives in shaping a sustainable future.

References

Abdelaziz C, Driss Z, Hassane H, Rania K, Mohammed H (2022) Agriculture recursive, Module SmartCode de la ville Tamansourt

Al Omrane (2010) Charte urbaine, architecturale et paysagère des villes Nouvelles marocaines

Al Omrane Marrakech (2008) Tamansourt, une ville nouvelle aux portes de Marrakech. p 181, 183

Alexander C (1977) A pattern language: towns, buildings, construction. Oxford University Press

City-facts (n.d.). Retrieved 20 November 2022, from https://www.city-facts.com/tamansourt-harbel-marrakesh-safi-morocco

Côté-Roy L, Moser S (2022) A kingdom of new cities: Morocco's national Villes Nouvelles strategy. Geoforum 131:27–38

Ecocity Builders (2010) What is an ecocity? https://ecocitybuilders.org/what-is-an-ecocity/

Ecocity Builders (2022) Feedback on plants/biodiversity from IUCN on Ecocity Morocco Jury. [e-mail]

Ecocity Builders (n.d.) EcoCity framework & standards. https://ecocitybuilders.org/ecocity-standards/

Ecocity Morocco | New Eco-Cities and Eco-Districts Across Morocco (n.d.) Retrieved 8 August 2022, from https://ecocitymorocco.org

Fniguire F, Laftouhi NE, Saidi ME, Markhi A (2014) Some aspects of climate variability and increasing aridity in central Morocco over the last forty years: case of Tensift Basin (Marrakech-Morocco). J Environ Earth Sci 4(9):42–51
Google Earth Pro (2022) Maps Tamansourt, Chrafate, Tamesna, Lakhyayta [Satellite images]. Retrieved September 19, 2022, from https://earth.google.com/
Hill, M., Pimentel, M. (2022). Morocco reflects a global South Dilemma: water or food? https://www.usip.org/publications/2022/06/morocco-reflects-global-south-dilemma-water-or-food
Keeton RE (2020) African new towns: an adaptive, principle-based approach. A+BE | Architecture and the Built Environment. https://doi.org/10.7480/abe.2020.18
Keeton RE, Provoost M (2019) To build a city in Africa: a history and a manual. nai010 publishers
Maja MM, Ayano SF (2021) The impact of population growth on natural resources and farmers' capacity to adapt to climate change in low-income countries. Earth Syst Environ 5(2):271–283
Nijhuis S (2022) Landscape-based urbanism: cultivating urban landscapes through design. In: Design for regenerative cities and landscapes. Springer, Cham, pp 249–277
Rajaud A, Noblet-Ducoudré ND (2017) Tropical semi-arid regions expanding over temperate latitudes under climate change. Clim Change 144(4):703–719
Ram M, Gerretsen P (2014) Veel ruimte! Maar durven we het ook te benutten?
Roggema R (ed) (2020) Nature-driven urbanism. Springer, Dordrecht/Heidelberg/London. 359pp
Stockholm Resilience Centre (2016) Contributions to agenda 2030. https://www.stockholmresilience.org/research/research-news/2017-02-28-contributions-to-agenda-2030.html
United Nations (UN) (2015) Transforming our world: the 2030 agenda for sustainable development
World Atlas of Desertification (n.d.) Patterns of aridity. https://wad.jrc.ec.europa.eu/patternsaridity

Chapter 8
Transforming Cities: The Impact of Small-Scale Green Interventions

Montserrat Flores, Catalina Cavazos, and Brenda Perez

Abstract A shared vision guided the development of Parque Brisas, transforming it into a park that not only mitigates and prevents localized flooding through rainwater harvesting but also serves as a vibrant community space. The park operates as a circular rainwater basin, collecting water from surrounding semi-permeable landscapes and channeling it to sunken, lushly planted areas. At its lowest point, a central plaza and amphitheater serve as the focal points, with meandering paths leading visitors through the park's rain gardens and play areas.

The park's design integrates multifunctional elements, such as an edge that functions as both a waterway and a traffic barrier, enhancing play and recreation opportunities. Semi-arid vegetation plays a crucial role in adapting to the region's climate, aiding in groundwater recharge, and supporting local biodiversity. This vegetation also contributes to temperature regulation and offers shade.

The revitalized space, previously underutilized, now becomes a dynamic urban area that invites daily enjoyment. Colorful concrete stepping stones and established trees encourage exploration and interaction with the green infrastructure.

Keywords Green infrastructure · Landscape design · Co-design · Stormwater management · Rain harvesting

8.1 Introduction

How can we develop infrastructure that adapts and evolves with the growing threat of climate change? How can disaster planning incorporate multi-purpose infrastructure without falling back on traditional approaches that address one problem while neglecting associated and resulting issues? Parque Brisas (Fig. 8.1) aims to create twenty-first-century infrastructure that mitigates risk for vulnerable neighborhoods,

M. Flores · C. Cavazos · B. Perez (✉)
PEW Studio, Monterrey, Mexico
e-mail: brenda@pew.mx

R. Roggema (ed.), *Dry Urbanism*, Contemporary Urban Design Thinking,
https://doi.org/10.1007/978-3-031-77347-1_8

Fig. 8.1 Overview of the site

restores ecosystems, and reconnects communities. While the project alone does not prevent district-wide flooding, it fosters the development of an ecologically beneficial and lower-risk project.

Public parks can do more than just host activities and recreation; they can capture, clean, and re-use stormwater.

At Parque Brisas, we designed a livable landscape that manages different types of water by layering infrastructure both above and below ground. The design holds stormwater runoff from the streets, parking lots, rooftops, other hard impervious surfaces neighboring the site, and expresses water at the surface in the form of inactive fountain during flooding events. This is then gradually released into the groundwater, and the atmosphere.

8.2 Location and Hydrological Context

Parque Brisas (Fig. 8.2) is located in Saltillo, the capital city of the state of Coahuila in Mexico. The city has diverse topography characterized by its location in the Chihuahuan Desert and its proximity to the Sierra Madre Oriental Mountain range. The city lies north of the foothills of this mountain range on a semi-arid plain. While the area is generally dry, there are occasional streams and rivers that flow through the region, especially during the rainy season.

The rapid growth of Saltillo's population has generated unrestrained urban sprawl, causing vegetation removal, extensive paving with artificial impermeable surfaces, elimination of natural water storage capacity and disruption of water flow paths. As a result, the city is especially vulnerable to changes in rainfall patterns due

Fig. 8.2 Site analysis

to the disruptions to natural systems. Urban flooding can have severe consequences on the environment, society, and the economy. Furthermore, it is expected that climate change exacerbates the frequency and intensity of precipitation events. The effects of increased hydrological events due to climate change could damage the social and ecological systems in urban areas.

Seasonal flooding events in Saltillo have shown that traditional stormwater management methods can't handle the water-related effects of climate change. Consequently, in recent years the advocacy to adopt green infrastructure strategies to manage stormwater runoff and flooding mitigation has risen sharply on several policy agendas within the municipality.

8.3 Research-Based Approach

We relied on a research-based approach using spatial data and analytical tools to examine and understand the urban environment, mapping and modeling, and on-site quantitative and qualitative data on micro-climate, soil conditions, existing vegetation, topography, drainage and human behavior (Fig. 8.3).

This method allowed us to better integrate findings into the planning and implementation stages to ensure that design decisions were informed by the latest knowledge and best practices.

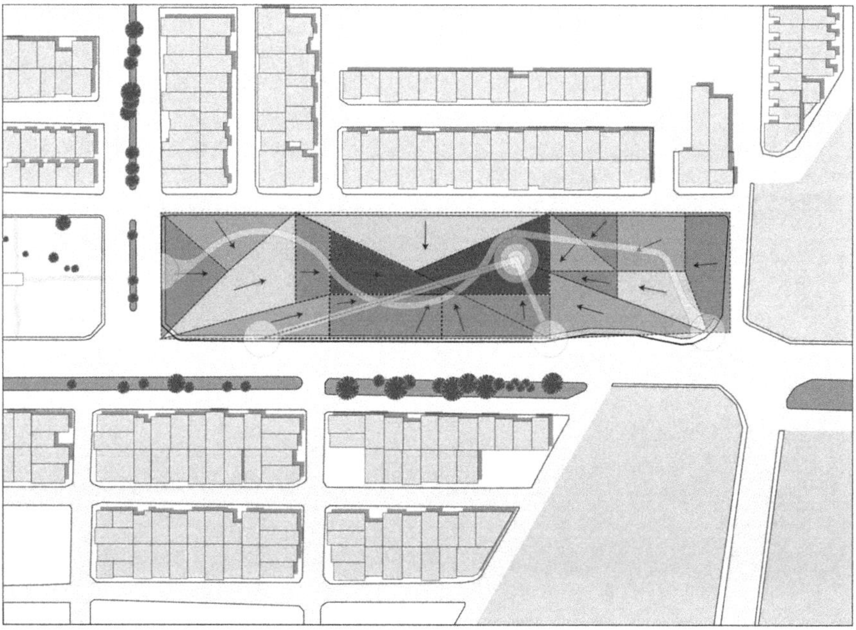

Fig. 8.3 Site Origami

Design Challenge

- Capture and clean stormwater from the surrounding residential roads (Fig. 8.4)
- Increase habitat and species diversity in order to improve overall resilience of the ecosystems in the park.
- Create more opportunities for physical activity, draw additional visitors to the park, and elevate public enjoyment.
- Enhance connectivity and circulation by upgrading the existing path system, stepping stones, and seating areas.
- Enrich the visual experience of the park by highlighting the "water story," fostering opportunities for green infrastructure learning.
- Strengthen the park's sense of place and character for surrounding neighborhoods by making the green infrastructure strategies a central feature of the park's attractive setting (Fig. 8.5).

8.4 Collaborative Design

Our process started by listening to the community. With the IMPLAN and local council we engaged in an open dialog that allowed us to understand the community's needs, aspirations and concerns. Blank maps of the project site where residents could mark or draw their desired park features were used along with an interactive

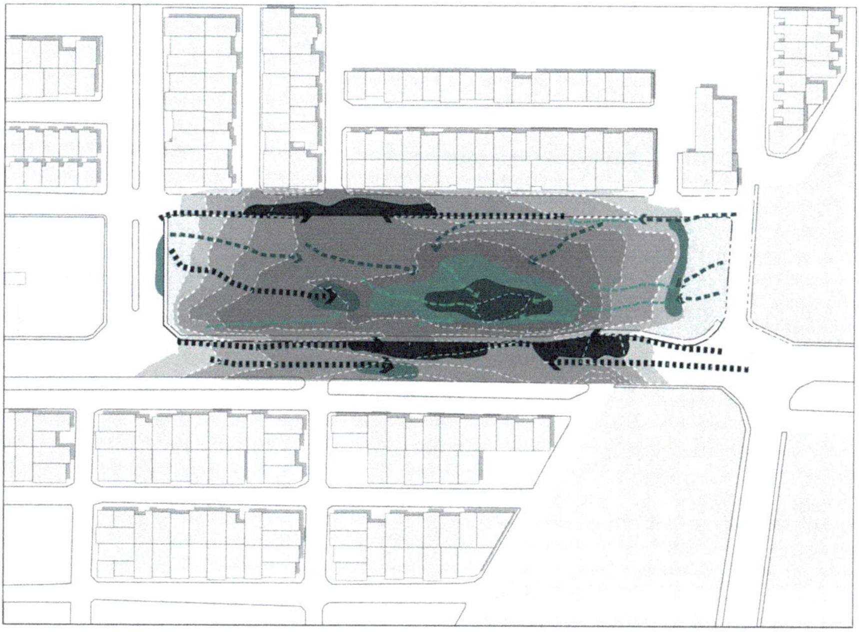

Fig. 8.4 Possible inundation patterns

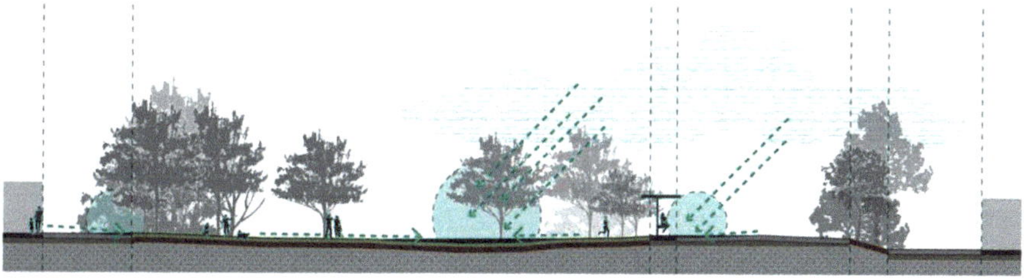

Fig. 8.5 Section

voting system with stickers to enable participants to view communal priorities in real time (Figs. 8.6 and 8.7).

By engaging in meaningful conversations, we gain deep insights into the community's vision for their ideal park creating common understanding of problems and needs. This approach established the foundation for our design that was both practical and purposeful.

A collaborative approach involving diverse expertise in the co-design process enabled us to focus on efficient design outcomes that could address various needs and personal-social situations.

After the design was finalized and approved by the community, we developed a comprehensive presentation that highlights the vision for the park. This presentation features 3D renderings, layout plans, and a narrative explaining how the

Fig. 8.6 Collaborative design work

Fig. 8.7 Design workshop

green-infrastructure elements aligned with the community's goals. We then share it with local stakeholders, including the city council, and neighbors, to generate support and build momentum for the project.

During the construction process, we keep open lines of communication with the stakeholders by providing regular updates and addressing any concerns that arise. Our close connection with the on-site team enables us to make adjustments as needed during construction.

8.5 Water-Based Vision

A shared vision emerged to create a park that not only mitigates and prevents localized flooding by harvesting rainwater but also serves as a public space for the community, showcasing a series of gardens that flow throughout the landscape.

8.5.1 Basin

The circle, a fundamental geometric shape frequently seen in nature and universally recognized, symbolizes various concepts including unity. The park works as a circular rainwater basin. At its perimeter, the water is collected from surrounding semi-permeable landscape and guided by the topography to sunken parts of the park covered in lush water planting (Fig. 8.8). At the lowest section, an amphitheater was proposed; a central plaza designed as the focal point of the park. The park is further articulated by meandering paths that take us across the basin and rain gardens, passing by benches and play elements (Fig. 8.9).

As a result, Parque Brisas has transformed into a vibrant green space that seamlessly blends urban nature, community, and climate adaptation in one. The challenges of managing rainwater have been turned into an opportunity, with rainwater now enhancing the area by creating engaging blue and green recreational spaces. The combination of flood protection with increased space for recreation and relaxation has long been a key aspiration of the residents.

8.5.2 Rain Garden

A garden bed runs longitudinally along the residential street, capturing rainwater through a water garden bioswale, conducting the sifted water to the larger basin (Fig. 8.10).

The multifunctional edge element framing the urban space acts both as a waterway and a barrier against traffic (Fig. 8.11). The space encourages play and recreation, inviting children and explorers to navigate narrow gravel paths, balance along

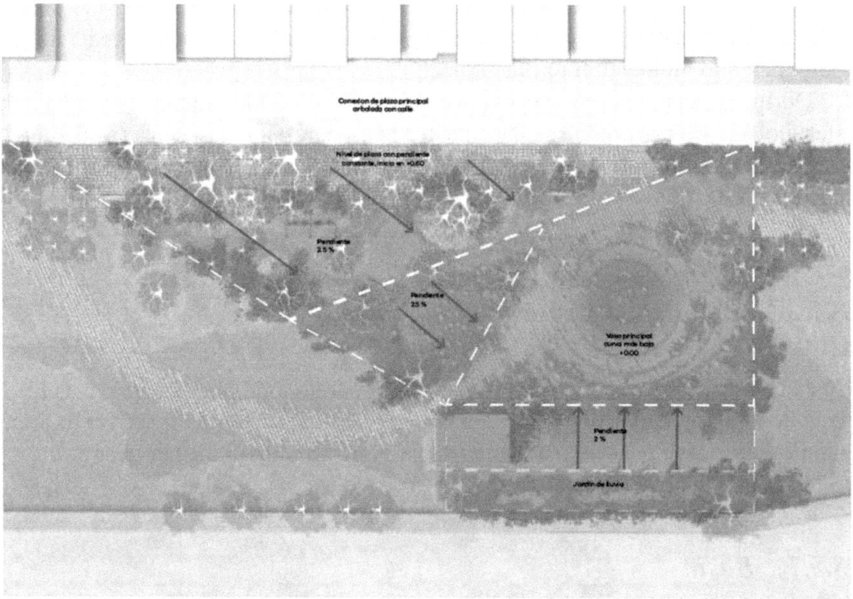

Fig. 8.8 Collection of rainwater

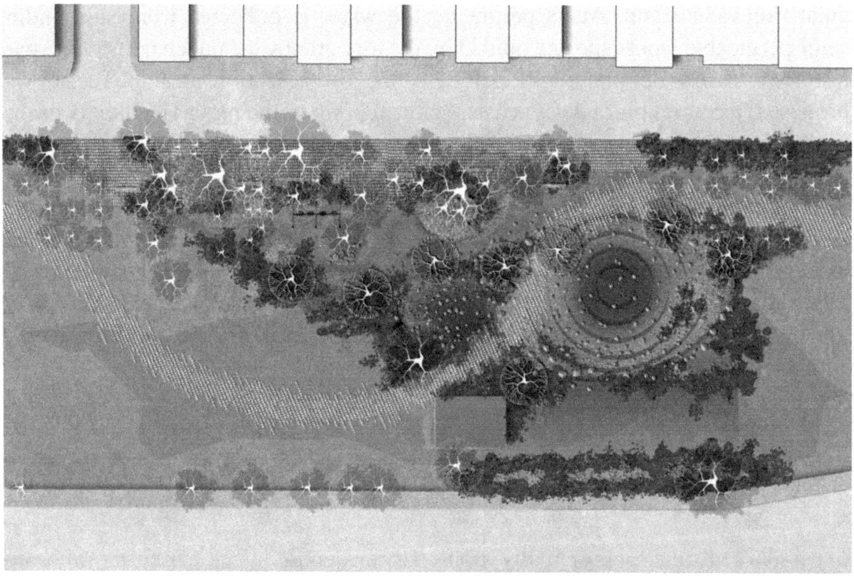

Fig. 8.9 The circle as the central point for water collection

Fig. 8.10 Construction of the central area

the colorful elements, or discover the garden rich in life and biodiversity. Benches are positioned to offer relaxation, foster friendly conversations with neighbors, and provide a peaceful retreat amidst the greenery.

The selected semi-arid vegetation plays an important role in the adaptability of the park to the region's climate change challenges (Fig. 8.12). These drought-tolerant plants, equipped with deep root systems, waxy leaves, and water-storing abilities, thrive in the hot, dry conditions typical of the area. By capturing rainfall, they aid in groundwater recharge and lessen the strain on stormwater management systems. Additionally, the rain garden vegetation supports local biodiversity and helps regulate temperatures, offering shade and cooling the urban environment.

The underutilized residual area has been transformed into a vibrant urban space for residents to enjoy daily. During rainfall, the green oasis becomes a picturesque water basin, enhancing the area's flood protection.

8.5.3 Paths and Playscapes

The newly planted rain garden and established existing trees invite residents to immerse themselves in the park exploring the green infrastructure elements, engaging them spatially. Within this setting colorful concrete "steppingstones" up to 60 cm height, become both path and play (Fig. 8.13).

Fig. 8.11 The longitudinal space along the road

The steppingstones are used as a landscape feature that encourages the users to connect with nature through play and experience a sense of personal adventure and discovery, rather than simply moving efficiently from one point to another.

Fig. 8.12 Central circle under construction

8.5.4 Educational Signage

Public outreach is a key element of a green infrastructure programs, as it helps build community support among government officials, and residents. Educating the

Fig. 8.13 Colorful stepping stones

community about green infrastructure can raise awareness of its benefits. Residents can play a role in effectively managing green infrastructure by clearing trash and debris and maintaining vegetation. The signs placed near green infrastructure can grab attention, inform people about these projects, and communicate their significance and purpose (Fig. 8.14).

Engaging and informative educational signage mark rain gardens, vegetated buffers, and other green stormwater infrastructure though the park. The signs in Parque Brisas serve as an educational tool that informs, connects people with water and nature, and inspires a desire to protect it.

Fig. 8.14 Educational signage

8.6 Lessons Learned

8.6.1 Promotion and Education of the Benefits of Green Infrastructure

The first recommendation is to acknowledge green infrastructure for its multiple benefits. Green infrastructure should be promoted as a strategy that delivers benefits to different stakeholders and meets multiple policies and strategic objectives across municipal departments.

Promoting green infrastructure's multi-functionality can help boost its implementation by involving stakeholders that benefit from these strategies, including communities. This aligns with respondents' concerns regarding the perceived lack of understanding from private developers, community and decision-makers around green infrastructure and its potential to tackle diverse urban challenges at the same time.

The value of green infrastructure (Fig. 8.15) depends on the environmental and social context; therefore, engaging with communities can help to develop a broader understanding of its benefits and values. Consequently, this might raise support for the implementation and ownership of green infrastructure projects. Active community engagement has greater capacity for cultural and behavioral change in comparison to merely depending on public observation of green infrastructure.

Fig. 8.15 Value of green infrastructure

8.6.2 Documentation, Monitoring, and Evaluation

After-completion the documentation and evaluation of local projects, regardless of scale, is imperative to promote and spread the use of green infrastructure. Even though green infrastructure projects are being developed in the city, there is no project database that contains information and details of each strategy such as: elements involved, geographical scale, sectors involved, costs and benefits (Fig. 8.16).

8.6.3 Cross-Sectoral Integration

The promotion of green infrastructure's multi-functionality at an institutional level can be further advanced by cross-sectoral integration.

Collaboration between different departments and organizations, as well as actively promoting partnerships between them, would help address the risk of ineffective communication and interagency fragmentation.

The collaboration between CityLab, Implan Saltillo, Saltillo City Council and private developers showed effective partnerships between water and flood management professionals.

Fig. 8.16 Essential monitoring of the rain garden

8.6.4 Innovative Funding Mechanisms for Small-Scale Projects

Even though no local funding mechanisms for green infrastructure are in place in Saltillo, financing options for the implementation of green infrastructure such as public-private partnerships, grants fees and taxes have been well documented in American cities by the EPA (2016).

The founding of the project was achieved through the collaborative efforts of the University of Stuttgart and the Fraunhofer Institute in Germany, along with key partners in Mexico, including the Municipal Planning Institute of Saltillo (IMPLAN) and the Monterrey Institute of Technology and Higher Education (ITESM).

Chapter 9
Integrating Dry Urbanism in Qatar's Urban Development: A Model for Sustainable Growth in Arid Regions

Hossam Ibrahim and Ibrahim Hegazy

Abstract This article presents an in-depth analysis of the integration of dry urbanism principles in Qatar's urban development, offering a comprehensive perspective on sustainable urban planning in arid regions. Dry urbanism, crucial for addressing the challenges of water scarcity and climate responsiveness, forms the core of Qatar's strategy to combat its harsh climatic conditions and rapid urbanization. The article begins with an overview of Qatar's geographic and climatic context, tracing the historical development patterns that have shaped its urban landscape. It then delves into the specifics of Qatar's development strategies, including the Qatar National Development Framework (QNDF), Climate Change Strategy (CCS) for Urban Planning, Qatar Urban Design Compendium (QUDC), and various other key policies and plans. These strategies are closely examined to highlight their alignment with essential aspects of dry urbanism, such as water management, energy-efficient urban design, and resilient infrastructure development. Through concrete case studies, the article showcases the practical realization of these principles in Qatar, evaluating their impact and effectiveness. It also discusses the challenges and opportunities inherent in implementing such strategies, positioning Qatar as a potential model for other arid regions facing similar challenges. The article concludes by emphasizing the critical role of integrating dry urbanism in urban planning for a sustainable future, particularly in arid environments like Qatar, and underscores the significance of stakeholder involvement for achieving these ambitious goals. This analysis not only sheds light on Qatar's approach to sustainable urban development but also contributes valuable insights to the global dialogue on adapting urban planning practices in the face of climate change and water scarcity.

H. Ibrahim (✉)
Department of Architecture, Mansoura Higher Institute for Engineering and Technology, Mansoura, Egypt

Faculty of Urban and Regional Planning, Cairo University, Cairo, Egypt

I. Hegazy
Department of Urban and Regional Planning, Faculty of Architecture and Planning, King Abdulaziz University, Jeddah, Saudi Arabia

Department of Architecture, Faculty of Engineering, Mansoura University, Mansoura, Egypt

R. Roggema (ed.), *Dry Urbanism*, Contemporary Urban Design Thinking,
https://doi.org/10.1007/978-3-031-77347-1_9

Keywords Qatar · Dry urbanism · Sustainable development · Arid climate · Urban planning · Water conservation · Energy efficiency · Resilient infrastructure · Qatar National Development Framework (QNDF) · Climate Change Strategy (CCS) · Lusail city · Msheireb downtown Doha · Urban sustainability · Environmental challenges

9.1 Introduction

As global urban development progresses, the integration of sustainability principles becomes imperative, particularly in regions experiencing arid climates. The concept of "dry urbanism" has emerged as a critical framework for sustainable development in areas where water scarcity and extreme temperatures are prevalent (Arandelovic and Vrsaljko 2020). This article examines Qatar, a nation distinguished by its arid conditions and rapid urbanization, serving as an experimental ground for applying dry urbanism principles. Qatar's journey toward sustainable urban development transcends its local context, offering a global case study. The Qatar National Development Framework (QNDF) and the Climate Change Strategy (CCS) for Urban Planning demonstrate Qatar's dedication to creating a resilient urban future (QNDF 2016; CCS 2017). These strategies, consistent with the core principles of dry urbanism, provide a blueprint for water conservation, energy-efficient design, and sustainable infrastructure, all essential in arid environments (Salat and Bourdic 2012).

Integrating these principles amid rapid urbanization presents both opportunities and challenges. On one hand, rapid growth provides a platform for innovative urban planning; on the other, it raises critical questions about resource management, environmental impact, and sustainable growth. This article begins by contextualizing Qatar's geographical and climatic conditions, offering insights into the historical development patterns that have shaped its urban landscape. It systematically examines key elements of Qatar's development strategies, highlighting their alignment with dry urbanism principles (QUDC 2019). Through this comprehensive analysis, the article underscores the importance of integrating dry urbanism in urban planning, particularly in arid regions like Qatar. The insights gained from Qatar's experience are valuable not only for similar arid environments but also for the broader discourse on sustainable urban development amid global environmental challenges (Shaka 2019; Zhou et al. 2024).

9.2 The Context of Qatar

9.2.1 Geographical Characteristics

Qatar, a peninsular Arab country, extends into the Persian Gulf from the Arabian Peninsula. Its geographical positioning endows it with unique climatic and environmental characteristics that significantly influence its urban development strategies. The country's terrain is predominantly flat and arid, characterized by sandy deserts and a coastline featuring salt flats, mangroves, and beaches. This landscape plays a crucial role in shaping the nation's approach to urban development, particularly through the lens of dry urbanism. The unique geographical features of Qatar necessitate urban planning strategies that withstand its harsh environmental conditions, focusing on sustainability and resilience (El-Aby 2017).

9.2.2 Climatic Conditions

Climatically, Qatar experiences long, sweltering summers with temperatures often exceeding 40 degrees Celsius, and relatively short, mild winters. The country's arid climate is marked by minimal rainfall, intense sun exposure, and high evaporation rates (Qatar Meteorology Department 2023). These conditions present a challenging environment for sustainable urban development, necessitating innovative approaches to manage heat and water scarcity (Arandelovic and Vrsaljko 2020). The extreme climatic conditions have driven the adoption of strategies such as green roofs, reflective building materials, and energy-efficient designs to reduce urban heat islands and enhance thermal comfort (Wheeler et al. 2019).

9.2.3 Historical Development

Historically, Qatar's urban development adapted to its desert environment, with traditional architecture reflecting the need to cope with the harsh climate. As shown in Fig. 9.1 buildings were designed with thick walls, narrow alleys were laid out to create shade, and wind towers provided natural ventilation (Salat and Bourdic 2012). The discovery of oil and gas reserves in the mid-twentieth century transformed the nation's economic landscape, leading to rapid urbanization and a shift in urban planning paradigms. This economic boom resulted in a surge of high-rise buildings and modern infrastructure in Doha, reflecting the country's new wealth but also introducing challenges related to sustainability and resource management (Al-Raouf 2016).

Fig. 9.1 Urban Fabric in old Doha 1978 (QUDC 2019)

9.2.4 The Impact of Rapid Development

This rapid development brought about significant changes in the urban fabric of Qatar. The modern skyline of Doha, with its array of high-rise buildings, is a testament to this transformation. Yet, this rapid urbanization also posed challenges, particularly in terms of sustainability in a region where natural resources are limited (QNDF 2016). The shift towards modern urban planning had to address these challenges by integrating principles of dry urbanism, such as water conservation, energy efficiency, and resilient infrastructure (Qatar Urban Design Compendium 2019).

The geographic and climatic context of Qatar provides a critical backdrop for understanding the nation's approach to urban development. In addressing its environmental challenges, Qatar has adopted the principles of dry urbanism to guide its urban planning strategies. These strategies, which will be explored in the subsequent sections, are tailored to address the unique needs of a region where water is scarce, temperatures are extreme, and sustainable development is essential for the future (Shaka 2019; Zhou et al. 2024).

9.3 Analysis of Qatar's Development Strategies

Qatar's urban development strategies, particularly in the context of dry urbanism, are guided by several strategic initiatives and policies designed to address the unique challenges of an arid environment while promoting sustainable growth. This section provides an overview of these key strategies and examines their alignment with dry urbanism principles.

9.3.1 Qatar National Development Framework (QNDF)

The Qatar National Development Framework (QNDF) is a comprehensive approach that reflects Qatar's commitment to sustainable urban development amidst its rapid urbanization. Figure 9.2 presented the Qatar national spatial strategy as per QNDF. Here's a detailed look at the key strategies outlined in the QNDF (2016):

- Sustainable Growth and Urban Design. QNDF places a strong emphasis on sustainable growth, focusing on creating compact urban designs. These designs aim to reduce the ecological footprint of urban areas, a principle that is especially pertinent in Qatar's arid climate. By promoting high-density urban areas, the framework seeks to limit excessive land consumption and reduce dependency on automobile transportation.
- Urban Growth Boundaries. To combat urban sprawl, the QNDF establishes urban growth boundaries. These boundaries are designed to contain the expansion of urban areas and ensure that development remains efficient and sustainable. By doing so, the QNDF encourages the development of mixed-use urban centers and transit-oriented developments. These areas are intended to foster a higher quality of urban life, with integrated residential, commercial, and recreational spaces that are accessible via public transportation, reducing the need for private car use.
- Green Infrastructure. The integration of green infrastructure is a pivotal aspect of the QNDF. It advocates for the creation of green belts and open spaces that incorporate natural landscapes into urban settings. This strategy not only enhances the aesthetic value of urban areas but also plays a crucial role in improving air quality, reducing urban heat islands, and providing recreational spaces for residents. Green infrastructure serves as a critical component in enhancing the ecological sustainability and livability of urban environments.

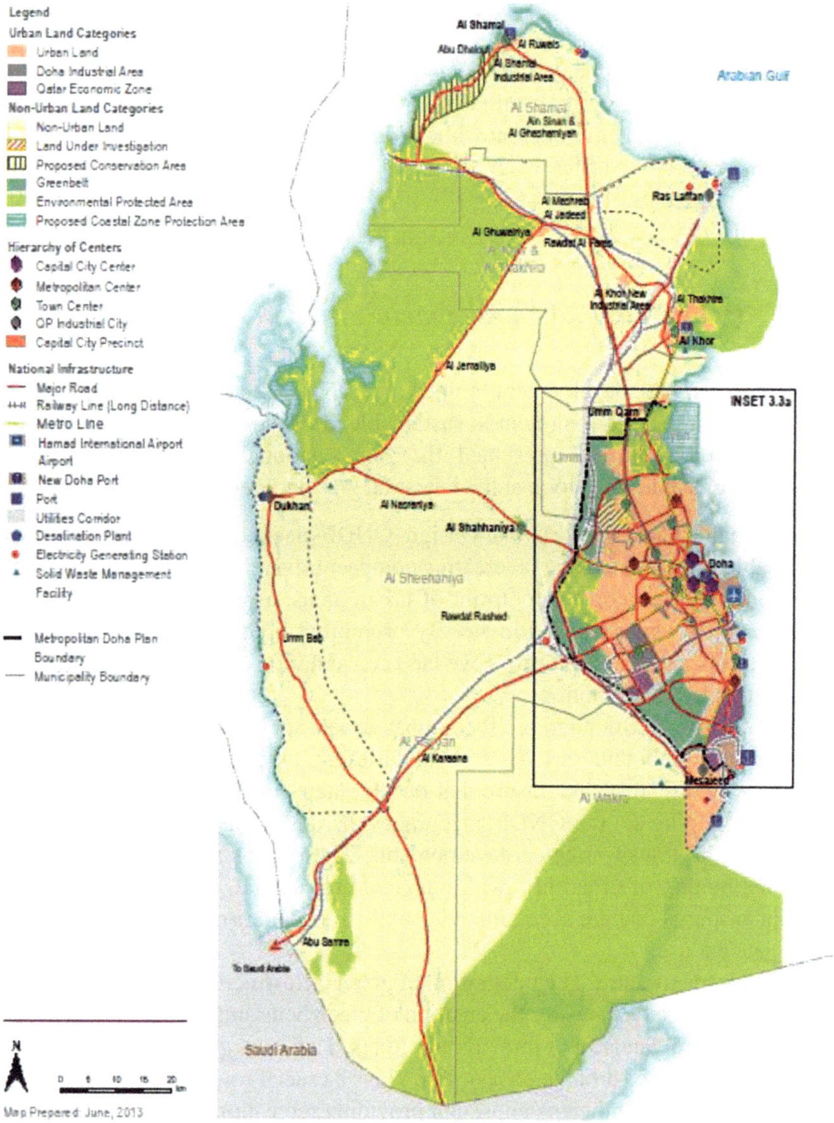

Fig. 9.2 Qatar national spatial strategy (QNDF 2016)

9.3.2 Climate Change Strategy for Urban Planning

The Climate Change Strategy (CCS) for Urban Planning in Qatar focuses on enhancing the resilience of urban areas to climate impacts, particularly vital in the country's arid environment. This strategy forms a crucial part of Qatar's broader commitment to sustainability and environmental stewardship (Salat and Bourdic 2012). Figure 9.3 presents the objectives of the Climate Change Strategy (CCS) for Urban Planning in Qatar. Here's a closer examination of the key components of this strategy:

- Reducing Urban Heat Island Effect. The strategy aims to mitigate the urban heat island effect, a common problem in densely built environments where built-up areas are significantly warmer than their rural surroundings. This is particularly crucial in Qatar, where high temperatures can exacerbate the discomfort in urban areas. Approaches such as increasing green spaces, using reflective building materials, and improving urban geometry to enhance airflow are some methods employed to combat this effect (CCS 2017).
- Enhancing Building Energy Efficiency. Central to the strategy is the emphasis on energy efficiency in buildings, which is critical for reducing the overall energy consumption and environmental impact of urban areas. The strategy promotes the use of energy-efficient designs and materials that minimize cooling demands. Techniques such as using reflective materials to deflect more sunlight and incorporating natural ventilation are encouraged to reduce the reliance on air conditioning. This not only helps in cutting down energy use but also in reducing greenhouse gas emissions associated with energy production from fossil fuels (CCS 2017).
- Encouraging Renewable Energy Integration. The integration of renewable energy sources into urban planning is a significant thrust of Qatar's climate change strategy. By incorporating technologies like solar panels into buildings and urban infrastructure, the strategy aims to decrease the dependency on fossil fuels. This shift is crucial for promoting sustainable energy consumption patterns and aligning with global efforts to combat climate change (CCS 2017).

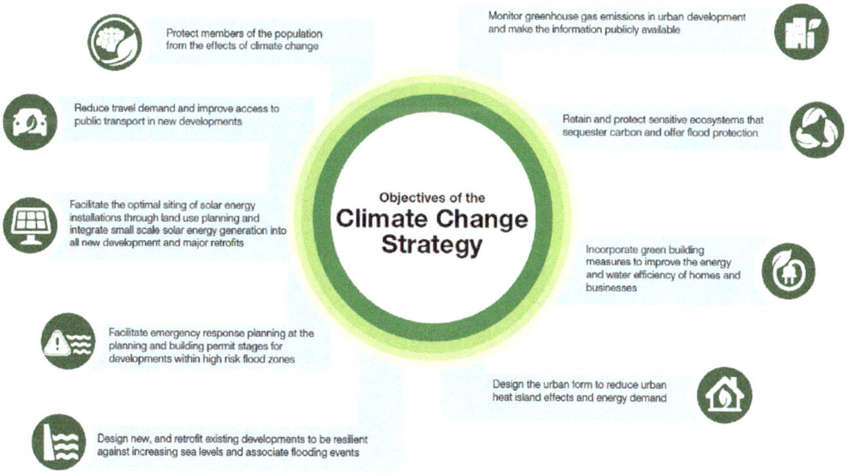

Fig. 9.3 The objectives of CCS for Urban Planning in Qatar (CCS 2017)

9.3.3 Qatar Urban Design Compendium (QUDC)

The QUDC is an essential guide for sustainable urban design tailored to the specific needs of Qatar's arid environment. It proposes innovative strategies to enhance urban livability while addressing environmental concerns. Figure 9.4 presents the QUDC components.

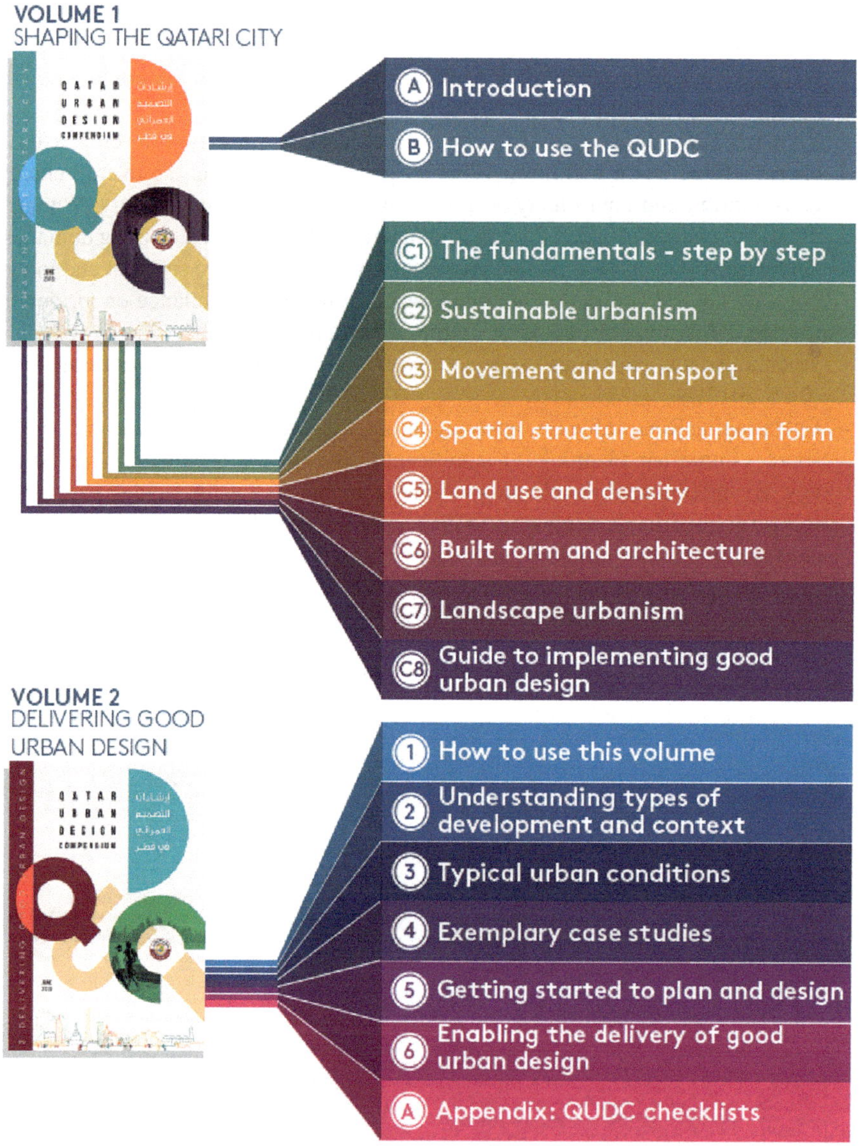

Fig. 9.4 The QUDC components (QUDC 2019)

The QUDC advocates for the integration of stormwater corridors into the urban fabric, serving dual purposes (QUDC 2019):

- Water Management. These corridors are strategically designed to manage stormwater effectively. By capturing and channeling runoff, they help mitigate flooding risks and improve water quality through natural filtration processes. This approach is crucial in urban areas where conventional infrastructure might struggle with sudden and intense rainfall.
- Recreational Spaces. Beyond their functional role in water management, these corridors are envisioned as green linear parks that provide valuable recreational and social spaces for the community. They enhance the aesthetic appeal of urban areas and contribute to the well-being of residents by offering accessible and inviting public spaces for leisure and social interactions.

9.3.4 Integrated Coastal Zone Management Plan (ICZMP)

The Integrated Coastal Zone Management Plan (ICZMP) is crucial for Qatar, given its extensive coastline and the environmental challenges associated with coastal development. Figure 9.5 presents delineation of ICZMP Study Area. The plan is designed to ensure sustainable and resilient coastal urban development by addressing several critical areas:

- Coastal Protection Zones. The ICZMP designates protected zones along the coastline to prevent erosion and habitat loss. These zones are essential for maintaining the structural integrity of the coast and preserving its natural ecosystems, which are vital for biodiversity and act as buffers against extreme weather events. Protecting these areas helps maintain ecological balance and supports the overall environmental health of the region (ICZMP 2015).
- Sustainable Coastal Development. The ICZMP also focuses on sustainable development projects that integrate environmental conservation with urban planning. An example of this is the Um Alhoul project, which features arid nature parks along the coastline. These parks not only help protect the coastline but also provide recreational spaces for the community, particularly those working in nearby industrial areas. This dual approach enhances the livability of coastal areas while ensuring that development activities do not adversely impact the coastal environment (ICZMP 2015).

9.3.5 Qatar National Housing Strategy (QNHS)

The Qatar National Housing Strategy (QNHS) is a comprehensive approach that underscores Qatar's dedication to sustainable urban development, particularly focusing on the unique challenges posed by its arid climate and scarce water

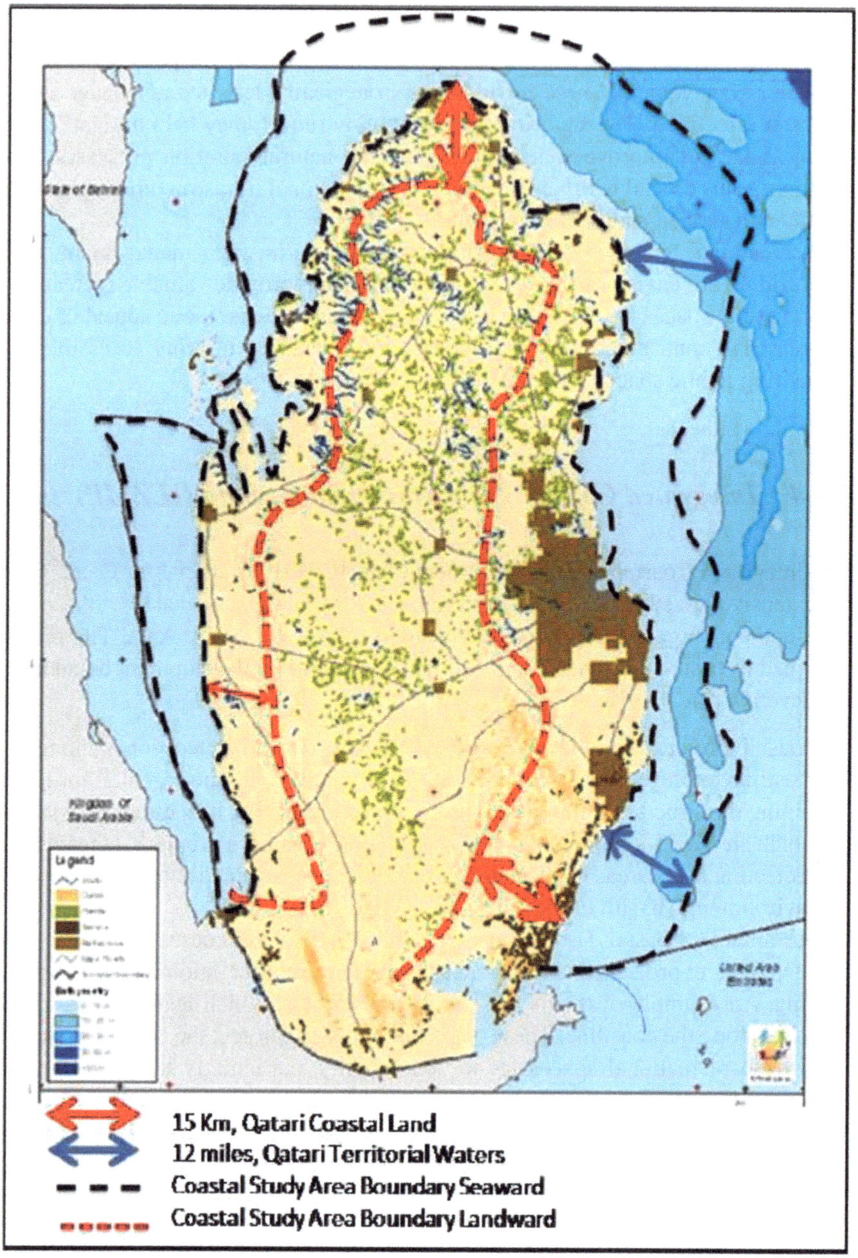

Fig. 9.5 Delineation of ICZMP Study Area (ICZMP 2015)

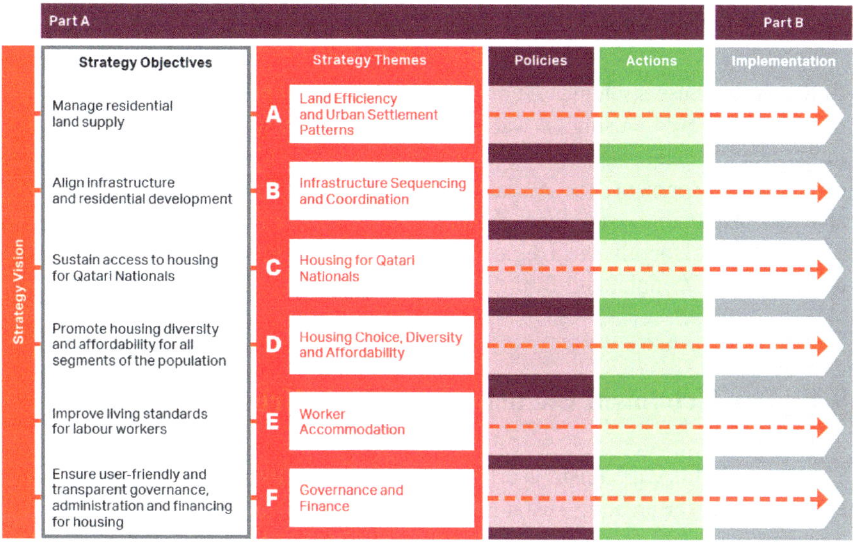

Fig. 9.6 QNHS: from vision to implementation. (QNHS 2020)

resources. Figure 9.6 presents QNHS: from vision to implementation. Here's a closer look at how the QNHS aims to address these challenges:

- Water-Efficient Housing. A key component of the QNHS is promoting water efficiency within residential developments. The strategy encourages the incorporation of water-efficient features, such as low-flow toilets and showerheads, which significantly reduce water usage in daily activities. Additionally, it advocates for the implementation of greywater recycling systems. These systems reuse water from sinks, showers, and washing machines for landscaping and flushing toilets, further cutting down the demand for fresh water. This focus on reducing water consumption is crucial in ensuring sustainable water management in an environment where water is a limited resource (QNHS 2020).
- Resilient Infrastructure. Recognizing the challenges posed by extreme weather conditions, such as high temperatures and occasional intense rainfall, the QNHS includes robust guidelines for constructing resilient infrastructure. These guidelines ensure that housing projects are not only sustainable but are also capable of withstanding adverse weather conditions. The focus on resilience helps in minimizing potential damages and disruptions, ensuring the long-term sustainability of housing developments (QNHS 2020).

9.3.6 Alignment with Dry Urbanism Principles

Collectively, these strategies demonstrate Qatar's alignment with the principles of dry urbanism. By focusing on water conservation, energy-efficient design, resilient infrastructure, and sustainable urban planning, Qatar addresses the core challenges of its arid environment. These strategies not only aim to make urban areas more livable and sustainable in the face of water scarcity and extreme temperatures but also serve as a model for other arid regions grappling with similar challenges (Arandelovic & Vrsaljko 2020; Salat and Bourdic 2012; Wheeler et al. 2019).

9.4 Dry Urbanism Elements in Qatar's Urban Planning

The concept of dry urbanism is crucial in arid regions like Qatar, involving various elements vital for creating sustainable and resilient urban environments. This section explores how Qatar has integrated these elements into its urban planning, demonstrating its commitment to sustainable urban development.

9.4.1 Water Conservation Techniques and Policies

In Qatar, water conservation is essential, and the nation has adopted various techniques and policies to ensure efficient water use. These include modern irrigation methods, such as drip irrigation in landscaping and agriculture, which has significantly reduced water wastage (QNDF 2016). The government has also invested in desalination plants to convert seawater into potable water, a critical resource in the desert climate (El-Aby 2017). Furthermore, there are policies in place to promote the recycling and reuse of wastewater for industrial and agricultural purposes, which has optimized Qatar's limited water resources (QNDF 2016). The implementation of policies to recycle and reuse wastewater in industrial and agricultural sectors is a key aspect of Qatar's water conservation strategy (QNDF 2016).

9.4.2 Climate-Responsive Architecture in Qatari Buildings

Qatar's architecture demonstrates a deep understanding of climate responsiveness. Modern buildings in Qatar are now being designed with a focus on energy efficiency and thermal comfort. This includes the use of high-reflectivity materials to reduce heat absorption, strategic building orientation to minimize sun exposure, and the incorporation of traditional architectural elements like courtyards and wind towers for natural cooling (QUDC 2019). One example is the Msheireb Downtown

Doha project, which uses narrow streets and dense massing for shading, as well as optimized building orientations to lower energy consumption (QUDC 2019).

9.4.3 Sustainable Infrastructure Development

Qatar's urban development strategy emphasizes sustainable infrastructure, focusing on the construction of green buildings that adhere to international sustainability standards. These efforts are complemented by the development of efficient public transportation systems designed to minimize carbon emissions and the strategic incorporation of renewable energy sources, notably solar power, into the energy framework (CCS 2017). This commitment to sustainable development is further highlighted by the creation of green spaces and parks, which not only enhance air quality but also provide natural cooling in urban locales (QUDC 2019).

In the realm of architecture, Qatar champions contemporary designs that are energy-efficient, utilizing materials that reflect heat and incorporating natural ventilation techniques to maintain cooler internal environments without excessive energy use (CCS 2019). The country has also made significant strides in enhancing public transportation, with the development of a metro system and extensive bus networks. These initiatives reduce reliance on private vehicles, thereby decreasing carbon emissions and promoting a more sustainable urban environment (QNDF 2016).

9.4.4 Urban Planning Approaches in Qatar's Context

Qatar has strategically adapted its urban planning methods to tackle its distinct environmental challenges, emphasizing the development of compact, integrated urban centers. This approach aims to curtail urban sprawl and reduce the ecological impact of development. A key aspect of the planning is enhancing connectivity, which facilitates easy access to amenities and reduces the necessity for extensive travel. Moreover, Qatar's urban strategies incorporate resilience to climate-related risks, such as heatwaves and scarce rainfall, ensuring that the urban infrastructure is robust enough to withstand and adapt to environmental fluctuations (QNDF 2016).

The focus on creating mixed-use developments and transit-oriented designs further streamlines the urban footprint and bolsters sustainability. Additionally, the Qatar National Housing Strategy outlines rigorous guidelines for building infrastructure that can endure extreme weather conditions, cementing the foundation for long-term sustainability and resilience in urban development. These comprehensive planning efforts are crucial for maintaining a sustainable and adaptable urban environment in Qatar (QNHS 2020).

Collectively, these strategies showcase Qatar's strategic alignment with dry urbanism principles. By focusing on water conservation, energy-efficient design, resilient infrastructure, and sustainable urban planning, Qatar addresses the core challenges of its arid environment. These strategies aim to make urban areas more livable and sustainable in the face of water scarcity and extreme temperatures and serve as a model for other arid regions facing similar challenges (Arandelovic and Vrsaljko 2020; Salat and Bourdic 2012; Wheeler et al. 2019).

9.5 Case Studies and Practical Implementation

This section provides a detailed exploration into the practical application of dry urbanism principles in Qatar, illustrating through key case studies how the country has integrated sustainability into its urban development. These case studies not only highlight Qatar's commitment to sustainable development but also offer insights into the effectiveness of these strategies and their broader implications for other arid regions.

9.5.1 Lusail City

Lusail City, strategically situated near Doha, is a flagship example of dry urbanism implemented on a large scale. This purpose-built city is designed to function as a sustainable and smart urban hub, incorporating advanced water conservation systems such as sophisticated wastewater treatment facilities that recycle water for various uses including irrigation (QNDF 2016). Figure 9.7 presents Lusail City Master Plan. The city's urban design is thoughtfully planned to minimize heat absorption; buildings and streets are oriented to maximize shade and promote natural air flow, enhancing natural ventilation throughout the city. Moreover, Lusail City places a strong emphasis on public transportation and pedestrian-friendly pathways, drastically reducing the dependence on personal vehicles, thereby decreasing urban congestion and environmental degradation (Al-Raouf 2016).

Lusail City's adherence to dry urbanism principles has yielded significant environmental and social benefits, making it a standout example of sustainable urban development. By implementing advanced water recycling systems, the city has dramatically reduced its freshwater demand, showcasing an effective model for water conservation that can be replicated in other arid regions. The strategic design of buildings and streets that optimize shade and airflow has not only reduced the city's energy consumption by minimizing the need for air conditioning but also improved the comfort and well-being of its residents (Programme-EN 2022). Additionally, Lusail City's emphasis on public transportation and pedestrian pathways has helped decrease vehicular traffic and pollution, fostering a healthier, more sustainable urban environment. These successes demonstrate the feasibility of large-scale sustainable urban planning and highlight the potential for similar developments globally to adopt these innovative practices.

Fig. 9.7 Lusail City Master Plan

9.5.2 Msheireb Downtown Doha

Msheireb Downtown Doha represents a transformative approach to urban regeneration, turning Doha's historical commercial district into a vibrant, sustainable urban community. This project effectively merges traditional Qatari architectural styles with modern sustainable design principles (QUDC 2019). Figure 9.8 presents Msheireb Downtown Doha Master Plan. The buildings within the district are predominantly LEED-certified, reflecting their high standards in energy and water efficiency. The urban layout prioritizes pedestrian pathways, providing ample shaded spaces and green areas which help mitigate the urban heat island effect and contribute to a cooler, more pleasant urban environment (Msheireb Properties 2021).

The transformation of Msheireb Downtown Doha into a sustainable community offers a compelling narrative on the integration of heritage conservation with modern sustainability practices. The project's success in significantly reducing energy

Fig. 9.8 Msheireb Downtown Doha (Msheireb Properties 2021)

consumption and improving thermal comfort in the district is a testament to the effectiveness of its climate-responsive design and LEED-certified buildings. The focus on pedestrian-friendly spaces not only enhances the livability of the area but also encourages a more active, socially engaging urban lifestyle. This approach has revitalized a historic part of Doha preserving its architectural legacy while infusing it with new life and advanced sustainability standards (Al-Raouf 2016). Msheireb Downtown Doha serves as a vital case study for urban regeneration, illustrating that sustainable development can respect and uplift historical contexts, providing a road-map for other cities with similar conservation and modernization goals.

9.5.3 The Pearl-Qatar

The Pearl-Qatar, a luxurious man-made island in Doha, is an example of high-end development that incorporates dry urbanism principles. The island includes advanced stormwater management systems and utilizes water-efficient landscaping techniques to ensure judicious use of resources (ICZMP 2015; Al-Amadi et al. 2016). Its buildings are designed to optimize energy efficiency, and the urban plan encourages walkability and the use of communal spaces, promoting a healthy, sustainable lifestyle among residents (Al-Raouf 2016).

The Pearl-Qatar represents a unique blend of luxury and sustainability, challenging the notion that high-end developments cannot be environmentally conscious. Figure 9.9 presents the map of The Pearl showing the districts. The integration of

Fig. 9.9 The Pearl districts (Al-Amadi et al. 2016)

sophisticated stormwater management and water-efficient landscaping in The Pearl-Qatar has minimized its ecological footprint, setting a benchmark for sustainable practices in luxury developments. The design principles that encourage energy efficiency and communal living spaces have not only reduced the operational costs but also enhanced the quality of life for residents, making sustainability an attractive feature in the luxury market (Muneerudeen et al. 2016). This project demonstrates how sustainable development principles can be effectively incorporated into upscale real estate, offering valuable lessons on balancing commercial success with environmental stewardship. The Pearl-Qatar provides a blueprint for future luxury developments worldwide, showing how they can achieve sustainability without compromising on the quality or exclusivity that attracts buyers.

These case studies from Qatar demonstrate the versatile application of urban planning and architectural strategies in creating sustainable, functional, and thriving urban spaces that effectively respond to the challenges posed by arid climates. They not only enhance the quality of life for their inhabitants but also establish benchmarks for other regions aiming to implement sustainable urban development practices in similarly challenging environments.

9.6 Challenges and Opportunities

Implementing dry urbanism principles in Qatar, though progressive, faces several challenges. These obstacles also present opportunities for innovation and advancement in sustainable urban planning.

There are several challenges for a swift implementation of Dry Urbanism in Qatar:

- Resource Limitations. Qatar's arid climate and limited natural water resources pose significant challenges. The reliance on desalination for water supply is energy-intensive and environmentally taxing (QNDF 2016). The energy demands and environmental impacts associated with desalination necessitate the exploration of more sustainable water management practices (El-Aby 2017).
- Rapid Urbanization. The rapid pace of urban development in Qatar sometimes outstrips the implementation of sustainable practices, leading to environmental stress and resource overconsumption. Ensuring urban growth aligns with sustainable principles requires careful planning and regulation (QNHS 2020). This includes developing policies that integrate sustainability into all levels of urban planning and construction (Al-Raouf 2016).
- Cultural and Behavioral Factors. Adapting traditional lifestyles and consumption patterns to more sustainable practices can be challenging. There is a need for public awareness and education to foster a culture of sustainability (QUDC 2019). Behavioral change initiatives and community engagement are essential for achieving long-term sustainability goals (CCS 2017).

- Economic Considerations. The initial cost of implementing sustainable technologies and infrastructure can be high, potentially hindering the adoption of dry urbanism practices. Investment in these technologies requires significant financial commitment, which can be a barrier (QNDF 2016). However, long-term economic benefits, such as reduced operational costs and improved environmental quality, can outweigh these initial expenses (Salat and Bourdic 2012).

Dry Urbanism offers also opportunities for growth and improvement:

- Innovation in Water Management. Qatar has the opportunity to be a global leader in innovative water conservation techniques, such as advanced wastewater recycling, rainwater harvesting, and efficient irrigation methods. These innovations can mitigate the environmental impact of desalination and enhance water sustainability (ICZMP 2015). By investing in cutting-edge water management technologies, Qatar can set a benchmark for other arid regions (Simon 2023).
- Sustainable Urban Design. The challenge of rapid urbanization can be met with creative urban design that integrates green spaces, energy-efficient buildings, and smart infrastructure, enhancing the quality of urban life. Projects like Msheireb Downtown Doha demonstrate how sustainable design can be successfully implemented in dense urban settings (QUDC 2019). Such projects serve as models for incorporating sustainability into urban regeneration (Muneerudeen et al. 2016).
- Public Engagement and Education. Raising awareness about the importance of sustainability can lead to behavioral changes. Engaging the community in sustainability initiatives can foster a culture of conservation and environmental responsibility. Public education campaigns and community-based projects can significantly enhance public participation in sustainable practices (CCS 2017). Collaborative efforts between government, private sectors, and the public are crucial for effective sustainability education (Hossam and Hegazy 2014).
- Investment in Renewable Energy. Leveraging Qatar's abundant solar energy can reduce the carbon footprint of desalination and other energy-intensive processes, making the implementation of dry urbanism principles more economically feasible. Investment in solar energy and other renewable sources can transform Qatar's energy landscape, aligning it with global sustainability goals (CCS 2017). The integration of renewable energy sources in urban infrastructure can significantly contribute to reducing greenhouse gas emissions (Realpe and Jabareen 2024).

9.7 Qatar as a Model for Other Arid Regions

Qatar's acclaimed pioneering efforts in integrating dry urbanism into its urban development offer valuable insights and lessons for other arid regions worldwide. This section explores how Qatar's approach can serve as a model, providing a comparative analysis with other arid regions.

Compared with other arid regions the following remarks can be made:

- Innovative Water Management. Qatar's advanced water management techniques, particularly in wastewater recycling and desalination, set it apart from many arid regions. The country's comprehensive water conservation strategies, such as the use of treated sewage effluent (TSE) for irrigation and industrial purposes, demonstrate effective management of scarce water resources (QNDF 2016). Countries facing similar water scarcity issues can learn from these practices to enhance their own water sustainability efforts (El-Aby 2017).
- Climate-Responsive Urban Design. Qatar's emphasis on climate-responsive architecture, incorporating traditional design elements with modern technology, offers a template for urban development in arid climates. Projects like Msheireb Downtown Doha, which use high-reflectivity materials, strategic building orientations, and natural ventilation techniques, provide practical examples of reducing heat absorption and improving thermal comfort (QUDC 2019). This contrasts with some arid regions where urban design does not adequately address extreme temperatures (Arandelovic and Vrsaljko 2020).
- Sustainable Infrastructure Development. Qatar's commitment to sustainable infrastructure, including green buildings and renewable energy integration, is more advanced compared to many arid regions where such initiatives are still nascent. The incorporation of solar energy in projects like Lusail City highlights the potential for renewable energy to support sustainable urban development (CCS 2017). The emphasis on sustainable infrastructure also includes efficient public transportation systems, which reduce carbon emissions and promote environmental health (QNDF 2016).
- Qatar's holistic urban planning approach balances rapid urbanization with environmental sustainability. The development of compact urban centers, integrated public transport systems, and green spaces ensures that urban growth does not come at the expense of ecological health (QNDF 2016). This can provide guidance to other arid regions grappling with similar urban development challenges, demonstrating that sustainable growth and environmental stewardship can coexist (Salat and Bourdic 2012).

Qatar is keen to act as a frontrunner in the region and undertakes several activities in this context:

- Showcasing the feasibility. Qatar demonstrates that integrating dry urbanism principles in an arid environment is not only possible but also beneficial for long-term sustainability. The successful implementation of these principles in projects like Lusail City and Msheireb Downtown Doha serves as an encouraging example for other nations in similar climatic conditions (QUDC 2019). These projects illustrate the practical benefits of sustainable urban design and the feasibility of such initiatives in harsh environments (Al-Raouf 2016).
- Driving force of regional collaboration. Qatar's initiatives can foster regional collaboration and knowledge sharing among arid countries. By serving as a role model, Qatar can spearhead a movement towards sustainable development in

arid regions. Initiatives like the Integrated Coastal Zone Management Plan (ICZMP) show how collaborative efforts can address shared environmental challenges (ICZMP 2015). Regional cooperation can lead to the development of standardized practices and shared resources, enhancing overall sustainability efforts (Hossam and Hegazy 2014).

- Qatar's experience contributes significantly to the global discourse on sustainable urban development in arid environments. By addressing the unique challenges of arid regions, Qatar adds valuable perspectives to the broader sustainability conversation. The country's success in integrating dry urbanism principles provides a model for others to follow, promoting a sustainable and resilient future for urban areas worldwide (Salat and Bourdic 2012). Qatar's innovative approaches and practical implementations offer valuable lessons that can be adapted and applied globally (Realpe and Jabareen 2024).

9.8 Conclusion

This analysis has shown that the integration of dry urbanism principles in Qatar's urban development is not merely a response to environmental imperatives but also a visionary step towards sustainable urban living in arid regions. The insights from Qatar's experience provide valuable lessons for global urban development, especially in areas with similar climatic challenges. The following conclusions can be drawn:

- The Importance of Integrating Dry Urbanism in Urban Planning. Qatar's case highlights the critical importance of incorporating dry urbanism principles in urban planning. This approach is essential in arid regions, where managing scarce water resources, mitigating extreme heat, and ensuring sustainable growth are paramount. By embracing dry urbanism, cities can become more resilient, livable, and adaptive to the challenges posed by arid environments (QNDF 2016). The strategic emphasis on water conservation, energy efficiency, and climate-responsive design underscores the necessity of these principles in addressing the unique challenges of arid urban environments (Arandelovic and Vrsaljko 2020).
- Achieving sustainable urban development in Qatar and other arid regions requires the active involvement of various stakeholders, including government authorities, urban planners, architects, developers, and the community. Collaboration among these stakeholders is crucial for implementing effective strategies, fostering innovation, and ensuring that urban development aligns with the needs and aspirations of the people it serves (QUDC 2019). Engaging the public through education and participatory planning processes can enhance the acceptance and effectiveness of sustainable urban practices (Climate Change Strategy for Urban Planning 2017).

Looking ahead, Qatar's urban planning is poised to continue evolving, with a focus on further integrating sustainability and resilience into its urban fabric. Continued investment in renewable energy, green infrastructure, and innovative water management will be key (CCS 2019). Additionally, fostering a culture of sustainability among residents and stakeholders will be crucial for the long-term success of these initiatives (Integrated Coastal Zone Management Plan 2015). The adoption of cutting-edge technologies and practices in urban planning will support Qatar's vision of becoming a leader in sustainable urban development (El-Aby 2017).

Qatar's journey offers a blueprint for other arid regions seeking to embark on a path of sustainable urban development. The strategies and lessons learned can be adapted to suit different contexts, contributing to a more sustainable and resilient future for urban areas globally. As the world grapples with the challenges of climate change and resource scarcity, the principles of dry urbanism exemplified by Qatar will become increasingly relevant, offering guidance and inspiration for sustainable urban development in arid regions and beyond (Salat and Bourdic 2012). The successful implementation of these principles in Qatar demonstrates the feasibility and benefits of sustainable urban practices, providing a model for other regions to follow (Realpe and Jabareen 2024).

References

Al-Amadi D et al (2016) Form and function in the Pearl-Qatar artificial island development. Proc Int Conf Contem Aff Archit Urban 5(1):2022

Al-Raouf AA (2016) Emerging knowledge-based urbanism as an urban strategy in Qatar. ICUP 2016 proceedings, Nis: November 2016

Arandelovic B, Vrsaljko A (2020) Dry urbanism: sustainable urban design for arid regions. J Sustain Dev 13(3):50–65

CCS (Climate Change Strategy) (2017) Ministry of Municipality and Environment, Qatar

El-Aby F (2017) Towards sustainable urban development in arid regions: Masdar City as a case study. J Al Azhar Univ Eng Sec 12(42):199–211

Hossam SI, Hegazy I (2014) Towards applying strategic environmental assessment in the Egyptian context. Asian Acad Res J Soc Sci Hum 1(21):325–336

ICZMP (Integrated Coastal Zone Management Plan) (2015) Ministry of Municipality and Environment, Qatar

Msheireb Properties (2021) Msheireb downtown Doha: the vision. Msheireb Properties Documentation

Muneerudeen A, Al Khani F, Furlan R (2016) Urban revitalization of public spaces in the Pearl in Qatar. Am J Sociol Res 6(1):1–9

Programme-EN (2022) Sustainable cities in the Middle East and North Africa. CAREP—CITERES/EMAM study day, Paris

Qatar Meteorology Department (2023) Climate of Qatar. Ministry of Municipality and Environment, Qatar

QNDF (Qatar National Development Framework) (2016) Ministry of Municipality and Environment, Qatar

QNHS (Qatar National Housing Strategy) (2020) Ministry of Municipality and Environment, Qatar

QUDC (Qatar Urban Design Compendium) (2019) Ministry of Municipality and Environment, Qatar

Realpe SN, Jabareen Y (2024) Categorization framework of adaptation measures in urban dry arid climate. EGU General Assembly 2024

Salat S, Bourdic L (2012) Urban complexity and sustainability. Springer

Shaka H (2019) Rethinking cities in arid environments for the 21st century. The Nature of Cities
Simon D (2023) Urban climate change governance and hot and arid zones: mutual learning and partnerships to build sustainable resilience solutions. Earthna Center for a Sustainable Future, Doha, Qatar
Wheeler SM et al (2019) Mitigating urban heating in dryland cities: a literature review. J Plan Lit 34(4):434–446
Zhou S, Nijhuis S, Dijkstra R (2024) Towards a pattern language for green space design in high density urban developments. J Urban Des. https://doi.org/10.1080/13574809.2023.2300505

Chapter 10
Projects for the Future City: A Design Framework for Climate Justice

Diana Garcia-Cejudo and Rodrigo Pantoja-Calderon

Abstract The purpose of this chapter is to recount the dichotomy of our city, Querétaro, a medium-sized city of one million inhabitants with a booming economy. Although the city is prosperous, it is also a victim of its success, facing environmental challenges due to urban and population growth. In addition, climatic conditions have accelerated and worsened due to climate change, exacerbating hydric stress and consequently demonstrating the limitations of existing infrastructure, putting the most vulnerable population at risk. To address the dichotomy and climate justice, the design studio 'Projects for the Future City,' conducted at the School of Architecture at Tecnológico de Monterrey, examines the climate justice mosaic of territory to explore the desirable possibilities of addressing multiple facets of local challenges. It addresses climate injustice, focusing on the socio-cultural and ecopolitical dimensions of the climate crisis, mainly in the city of Queretaro.

Through academic speculation, the design projects explore the possibilities and effects of different actions in response to the climate crisis's present and future urban challenges. Through a selection of academic projects that respond to specific climate justice, we reflect on how operating in an arid territory requires adaptation, resilience, and a deeper understanding of the opportunities and challenges of *dry urbanism*.

Keywords Climate justice · Research by design · Design studio · Dry urbanism · Global south

---

D. Garcia-Cejudo (✉) · R. Pantoja-Calderon
Tecnologico de Monterrey, Queretaro, Qro, Mexico
e-mail: dgarciace@tec.mx

R. Roggema (ed.), *Dry Urbanism*, Contemporary Urban Design Thinking,
https://doi.org/10.1007/978-3-031-77347-1_10

10.1 Introduction

The purpose of this chapter is to recount the dichotomy of our city, Querétaro, a medium-sized city of one million inhabitants in a booming economy. It is a prosperous city, part of a competitive region with significant economic growth due to private and foreign investment. Its economic development has been inextricably linked to its strategic location and the availability and access to resources: land and water. Since its foundation, the city has developed due to the available water sources and proximity to surface water bodies, historically supporting industrial production and urban consumption.

However, the city is also a victim of its success, facing environmental challenges from urban and population growth. Overusing natural resources such as water and land creates a polluted, unequal, and, for many, unaffordable area, which results in a depleted city. Moreover, sometimes prosperity only translates into high incomes for few residents with access to essential services and natural resources, creating more significant inequalities and areas of high vulnerability.

Another major challenge in this dichotomy is related to its context, which is climatic and natural conditions. Querétaro is characterized by a semiarid climate, which means high temperatures in summer and low temperatures in winter, with a short but intense rainfall period. "Within Querétaro, the minimum daily temperature increased to 0.751 °C per decade during 1982–2011, while the population increased by 33% during the last two decades (Colunga et al. 2015). These climatic conditions have accelerated and worsened due to climate change, exacerbating hydric stress and consequently demonstrating the limitations of the existing infrastructure, putting the most vulnerable population at risk. Operating in a dry territory requires adaptation, resilience, and a deeper understanding of the opportunities and challenges of *dry urbanism*.

"In the last decades, we have witnessed natural disasters and extreme weather events besieging human settlements across various continents at greater frequency due to climate change." (Carracedo García-Villalba 2021b, p. 121). Climate change disproportionately affects underprivileged and marginalized groups both globally and within Mexico.

Climate justice emphasizes climate change's differential, uneven, and disproportionate impacts. To address the dichotomy and climate justice, the design studio 'Projects for the Future City' examines the climate justice mosaic on a territory to explore the desirable possibilities of addressing multiple facets of the local challenges. Through academic speculation, the design projects explore the opportunities and effects of various actions in response to the climate crisis's present and future urban challenges.

This chapter builds upon the experience of the Design Studio conducted at the Architecture School at Tecnológico de Monterrey. This studio involves multidisciplinary teams, including partner universities, researchers, NGOs, community-based organizations, professionals, and stakeholders, collaboratively tackling the complex task of intertwining architecture, urbanism, and landscape design. The design studio addresses climate injustice, focusing on the sociocultural and ecopolitical dimensions of the climate crisis, mainly within the context of Queretaro City.

The chapter consists of four sections. The first section (10.1) contextualizes the problem by identifying the current challenges that open opportunities for resilience by defining climate justice design themes. Once the main challenges and opportunities are identified, the second part (10.2) develops the applied methodology, and later (10.3) through a selection of academic projects that respond to a specific climate justice design theme. Finally, some reflections and recommendations on the issues raised are presented.

10.2 Projects for the Future City: Challenges

The effects of climate change are evident worldwide, yet its impacts are not distributed equally. Both globally and within Mexico, climate change disproportionately affects underprivileged and marginalized groups. Climate justice emphasizes climate change's differential, uneven, and disproportionate impacts.

Regions with development constraints, such as Africa, Asia, Central and South America, LDCs, Small Islands, and the Arctic, are highly vulnerable to climatic hazards (Calvin et al. 2023). Increasing weather and climate extreme events have exposed millions of people to food insecurity and reduced water security, especially in locations and communities in the Global South. According to the IPCC, approximately 3.3–3.6 billion people around the globe live in contexts that are highly vulnerable to climate change. Between 2010 and 2020, human mortality from floods, droughts, and storms was 15 times higher in highly vulnerable regions compared to regions with very low vulnerability (Calvin et al. 2023). The severity of conditions is threatening communities and shaping their future.

To outline the local challenges, we revised different planning documents that identified the challenges and future paths for the city. The most recent Plan 2050, prepared by the Municipal Planning Institute (IMPLAN), identified five significant local challenges that shape the current situation and define the city's future challenges (IMPLAN 2020).

Land use planning or an urban management model cannot respond to or even guide the city's expansion. The existing planning documents and efforts are reactive. They react rather than create, control, or guide the city's development. "The city's growth criteria do not respond to a land use planning and urban governance model, resulting in a fragmented, discontinuous and segregated territory, evidenced in accelerated land use changes that seem to exceed the mechanisms for sustainable urban planning" (IMPLAN,2020). Additionally, the city does not have the mechanisms or sufficient regulations for strategic urban planning necessary to sustain the demands of the city dynamics.

Furthermore, economic growth is grounded on foreign investment and not founded on local economies. It causes fragilities in maintaining the local productive base and segregation in the regional workforce, especially among the unqualified and those with limited access to jobs, increasing social inequalities.

Unequal access to urban infrastructure, public services, and public space causes social inequalities and fragmentation. "These inequalities are emphasized in marginalized populations where situations of overcrowding and poor accessibility are perceived, situations that are contrasted with the proliferation of condominiums and gated communities with a wide oversupply of housing complexes in other areas of the city. This results in an increase in generalized violence and the perception of insecurity, which is detrimental to people's quality of life." (IMPLAN 2020).

The city's growth has essential implications for natural cycles and resources. This implication is especially true in terms of water management. "This is evident aided by the insufficient regulations of industrial water consumption and deficiencies in wastewater management. On the other hand, urbanization exponentially consumes land belonging to rural and environmentally important areas, increasing the risk of flooding, compromising the provision of ecosystem services, and, in general, increasing the risks associated with climate change" (IMPLAN 2020). Water is not distributed equally. The population with the least amount of harm is the most affected by the problem. Until now, accessibility has been a matter of privilege. The ever-present underregulated floods, migrations, outdated culture of consumption, and uncontrolled sprawl led us to this situation. The ongoing rate of consumption and toxic production that we lead as humans, the augmenting evaporation, and water demand (an increasingly scarce resource), combined with the lack of reparatory action, lead to present serious consequences.

The map below (Fig. 10.1) shows the city's expansion over the territory's natural conditions, demonstrating the challenging exploitation of natural resources such as fertile soil, water recharge areas, and groundwater.

Local and global challenges set a complex scenario that requires a deeper understanding of the challenges, opportunities, and responsibilities that act and guide development in the territory. Trying to act on this complexity, each of these local and global challenges is discussed through climate justice themes as a starting point to dissect the complex layers of the problem and explore the possibilities for the future.

10.3 Climate Justice Design Themes

Projects for the Future City aims to address climate justice by exploring the sociocultural and ecopolitical dimensions of the climate crisis. The studio is interested in contemporary problems that affect and transform the landscape, questioning the current practices that make the city, reassessing the work of our profession, and focusing on public space and social character.

These problems are affected by two aspects: their complex context and their multi-scalarity. Urban dynamics and the issues resulting from these respond to a complex context in physical, social, and demographic terms and whose impact or manifestation can occur at various scales (dynamic scale).

To address the complexity of the challenges described above, we defined a set of topics or lenses to set a starting point for researching and exploring the sociocultural

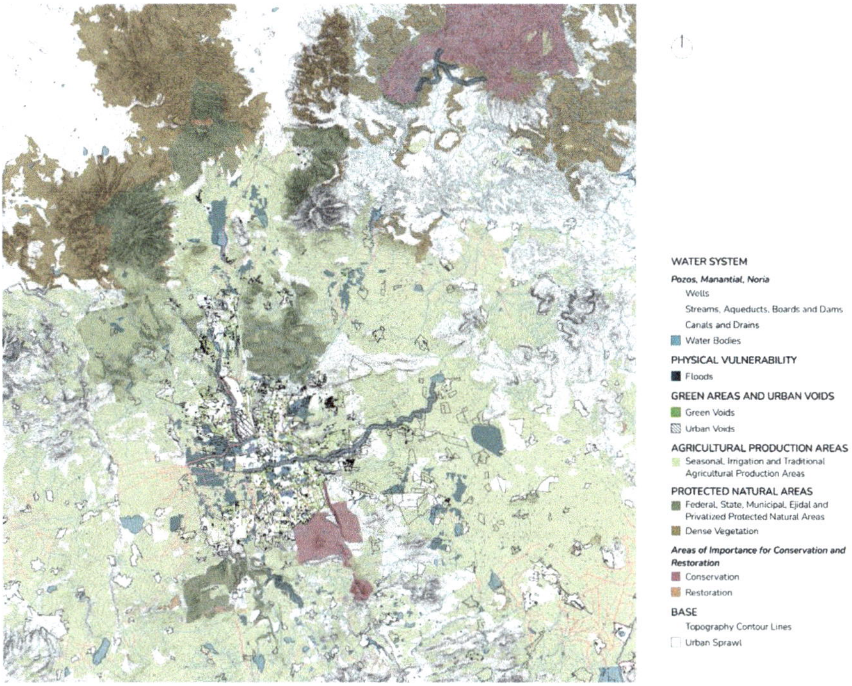

Fig. 10.1 Natural systems and urban expansion. Scale 1:400,000. (Source: Map made by authors based on data recovered from INEGI SRC WGS 84 UTM ZONE 14N 2023)

and ecopolitical dimensions of the climate crisis. Through four different *climate justice design themes*: hydric stress, energy transition, urban heat island, and soil degradation, the studio examines the climate justice mosaic on a territory to explore the desirable possibilities addressing multiple facets of the local challenges (Fig. 10.2).

Climate justice design themes are the starting point for academic speculation. Using the "research by design" methodology, the studio identifies local challenges resulting from the climate crisis, explores possible pathways and creates spaces of possibility where desirable futures are imagined, proposes an agenda that allows framing projects from a critical perspective and designs alternative narratives to a local practice dominated by short-term thinking.

This could be translated into a four steps methodology (Fig. 10.3) based on:

1. Agenda. Research on the socio-spatial, economic, environmental, and geopolitical dimensions that determine the agenda of climate injustice through the lens of each topic (hydric stress, energy consumption, waste management, and urban heat island).
2. Territorialization. Evidence on our socio-spatial, economic, environmental, and geopolitical territory. Identify the uneven effects of the climate crisis and how they interconnect. This phase includes the recognition of stakeholders and the normative dimension involved in each topic.

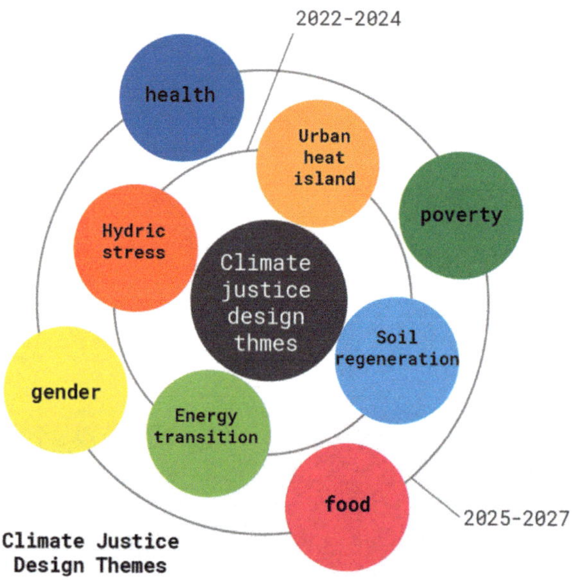

Fig. 10.2 Definition of Climate Justice Design themes by the design studio during 2022–2024 and future themes 2025–2027

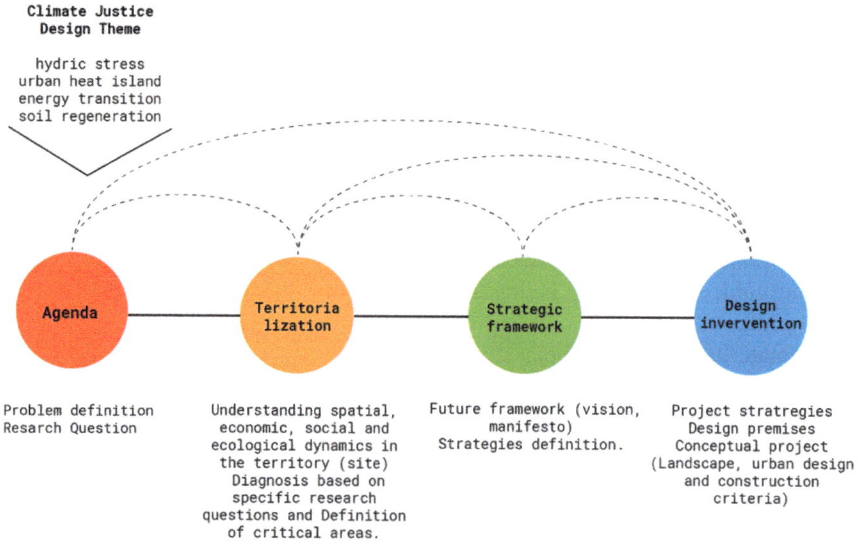

Fig. 10.3 Four steps design method

3. Strategic framework. Vision and spatial strategies for mitigation and adaptation identifying a set of critical actions, timeframe, and actors involved in the process
4. Design Intervention. A design project supported by the strategic framework that achieves a specific agenda topic in our territory.

The objective is to have a complete agenda, strategic framework, and design projects responding to the climate design themes to construct our city's desirable future.

10.3.1 Hydric Stress

The population growth resulting from the expansion of the urban sprawl has aggravated the city's water conditions. The supply of this resource is vital for the city's inhabitants, not only for living but also as a basis for economic production. In addition to the increasingly extreme climatic conditions, the city has an increasingly incipient water stress. "The Querétaro Valley is a region of large agro-industrial development that has allowed for an increase in population density and, with it, a change in land use and the availability of water resources. Despite the five aquifers in the valley of Querétaro, there is a reduction in groundwater levels of three meters per year. In the case of the Querétaro aquifer, urban consumption is the major demand for groundwater resources" (González-Sosa et al. 2013).

Furthermore, in an expanding city with an economy based on the overexploitation of natural resources, time is limited to act toward more resilient and sustainable water management. Following this aim, the studio focused on how architecture and landscape can contribute to transitioning to a water-sensitive city. On different scales, the academic projects deal with various challenges, such as the restoration of the Rio Querétaro, the regeneration of Rio Pueblito, and other green infrastructure that promotes the regeneration of the water bodies.

The first project is *Pieces of Time* (2021) by Natalia Olivera and Jimena Amieva. The project seeks to understand and make visible the current state of the water system in the Querétaro region, focusing on its main water body, the Querétaro River.. Placed on the northeast part of the river, the project aims to promote responsible, resilient urban development capable of creating a transition between the natural environment and the urban artifact, using the landscape as a medium. Through Nature-Based Solutions (NBS), the project proposes areas for ecosystem restoration, recreational spaces, and rainwater infiltration (Fig. 10.4).

10.3.2 Urban Heat Island (UHI)

The Urban Heat Island (UHI) effect is a human-caused weather modification. This phenomenon occurs when urban areas have higher temperatures than rural areas due to human activities such as construction, pavement, heat-absorbing materials, the lack of vegetation, and heat generated by vehicles and cooling systems.

This phenomenon represents a growing challenge for Latin American cities, exacerbated by rapid urban growth and the loss of permeable soil. "One of the best-known effects of urbanization is the increase in local heat. Modifying microclimatic conditions, especially in summer and places characterized by high temperature, causes comfort problems in outdoor and indoor spaces, increasing the energy demand for cooling" (Tumini et al. 2016).

In these terms, we question how architecture and landscape architecture can mitigate the UHI effect and create healthy environments for all. This studio analyzed a mixed-use urban area around the University campus. The study focused

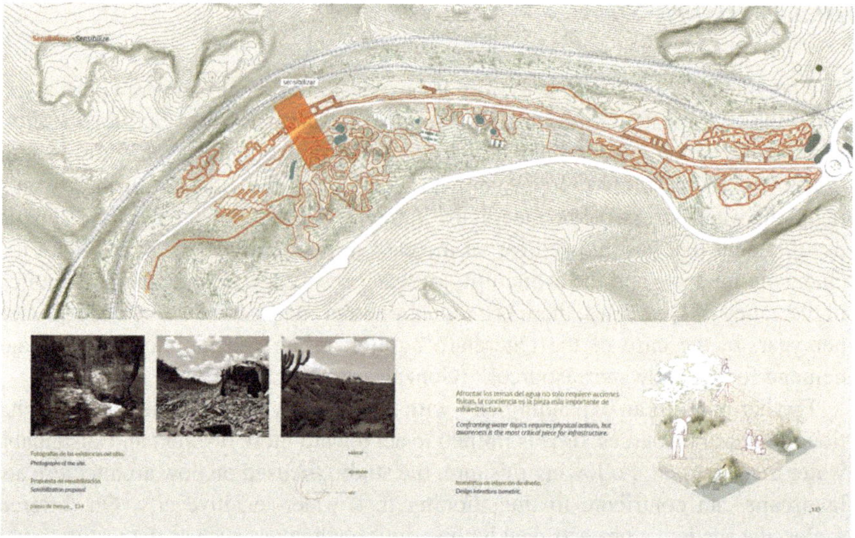

Fig. 10.4 Spatial Strategy Implementation

on temperature differences among diverse built environments, demonstrating that impermeable surfaces retain more heat and emphasizing critical areas for possible design proposals.

In this case, the design studio focused on a polygon around the University Campus of Tecnológico de Monterrey in Querétaro. This polygon has a radius of 1 km, and the purpose is to study the context and take temperatures of different spaces built with different materials. A combination of land uses can be observed within this radius, ranging from residential and industrial to educational, health, and commercial facilities. Consequently, typologies, densities, heights, and heterogeneous construction materials are diverse.

The first speculative academic project developed in our design studio is *Vía Verde*, designed by David Bolaños and Regina Macías (Macías & Bolaños 2023). This project is proposed to be in the area with the highest temperatures of the studied polygon, a monofunctional parking space of 50,000 square meters. In addition to rising temperatures, the parking area is not porous and does not offer any ecosystemic benefits. The "Vía Verde Project" renaturalizes the parking lot as a cooling point because green areas are fundamental in mitigating the heat island effect (Fig. 10.5). It also helps improve air quality through the evaporation of water in soils and the transpiration of plants. *Via Verde* aims to create a recreational space that expands the Querétaro 2000 Park, promoting wellness through a cool microclimate. It also utilizes "nature-based solutions" to mitigate flooding, capture and purify water, and foster biodiversity through pollinator gardens with native species.

The second academic project, *Cities of Care*, designed by Denisse Huerta and Ana Suárez, addresses the urban heat island effect by reusing a factory with high soil permeability and flexible infrastructure (Huerta and Suarez 2019). This site,

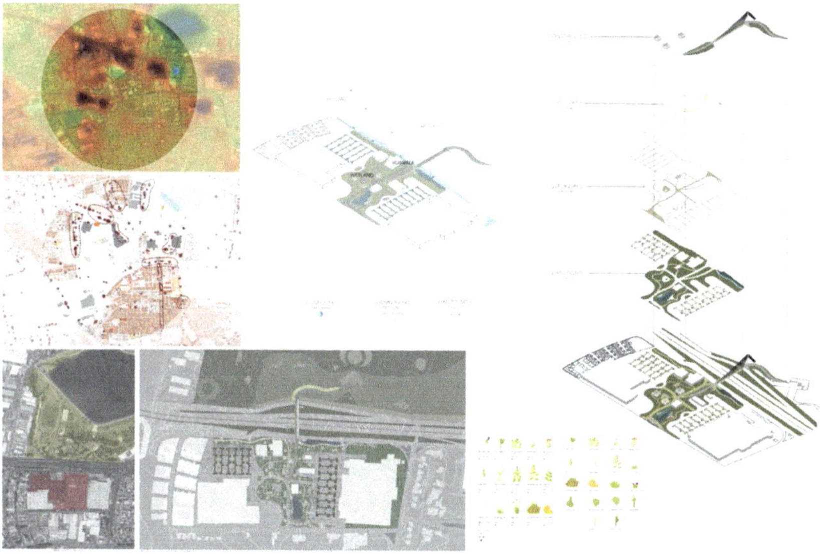

Fig. 10.5 Via Verde proposal placed on a vulnerable area (UHI)

chosen for its potential to mitigate heat, will concentrate various programs and reduce the need for long commutes. Landscape design plays a critical role in regulating temperature and restoring the ecosystem. It incorporates green spaces, wetlands, and productive landscapes to create a cooler environment (Fig. 10.6). In addition, the project includes spaces such as community kitchens, workshops, and a climate shelter to support the well-being of caregivers and the most vulnerable.

10.3.3 Energy Transition

Energy transition aims to shift from non-renewable, polluting energy sources to more sustainable ones, reducing greenhouse gas emissions and dependence on non-renewable resources. This studio examines how architecture and landscape architecture can aid in decarbonizing cities and fostering resilient environments. We focus on envisioning a desirable future and providing critical responses for the Vallejo Energy District in Mexico City. This part of the studio focuses on an industrial area in transition that plays an essential role in the CDMX economy.

Industrial Vallejo is CDMX's central industrial park, located within the northern Metropolitan Zone ring; it stands as a manufacturing titan on a national and industrial scale while also being engulfed by the growth of urban sprawl for many decades. These 500 hectares were previously divided into farmland and livestock ranches as part of Don Antonio Vallejo's hacienda, but in 1944, the industry was

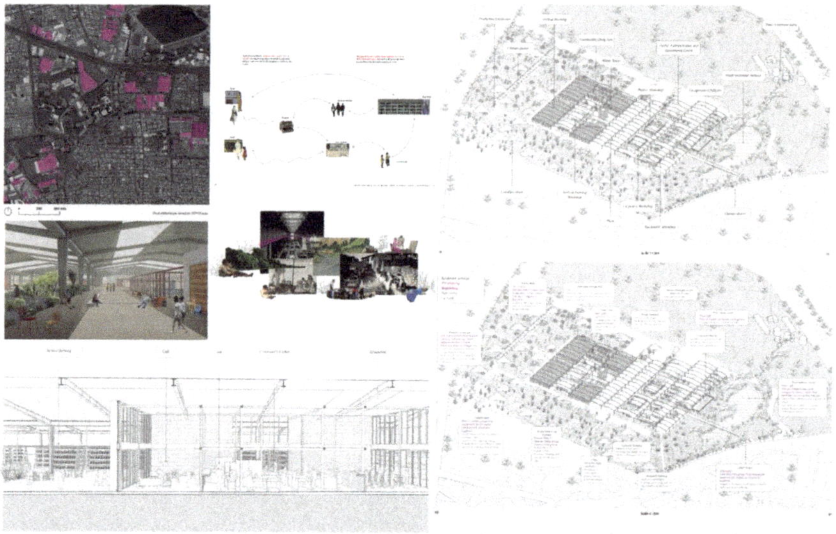

Fig. 10.6 Cities of care provide spaces for care for vulnerable populations, providing spaces and programs to reduce the long commutes and mitigate the UHI effects

declared the primary land use via presidential decree. If Vallejo is today dominated by industries, how can it spearhead CDMX's energetic transition through human-scale project insertions and efficient systems replicability?

As part of the research, it was discovered that there are many urban voids in Vallejo's industrial area. These voids, resulting from industrial decline, present unique opportunities for sustainable redevelopment by addressing Vallejo's contribution to GDP, polluted areas, lack of services, and the need to transition to cleaner energy sources to support the neighborhood's decarbonization.

The first project is PUNTO, by Alesia Armas, Valeria Márquez, and Diego Ramirez (Armas Sámano et al. 2024). It intertwines architecture, urbanism, and landscape architecture by revitalizing its site through public green spaces, housing, and clean energies (Fig. 10.7). With an ecosystemic approach, the new housing aims to create a sense of belonging and collective appropriation.

The second design proposal is CAPA, made by Santiago Altamirano Gama, Jimena Muñoz Chicano, and Sergio Yael Hernández Hernández. (Altamirano et al. 2024) The *Environmental Learning and Production Center* proposes to decarbonize Vallejo through the intersection of energy, architecture, and landscape architecture through programmatic strategies, with training spaces, daycare, and urban agriculture, all under a pre-existing industrial structure with a self-sufficient photovoltaic system (Fig. 10.8).

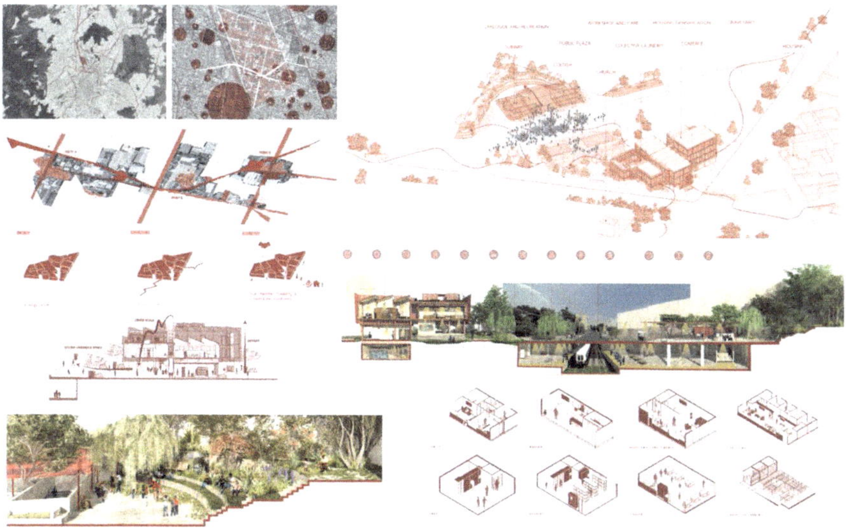

Fig. 10.7 Punto aims to provide affordable housing and a vibrant public space that generates green energy, contributing to the neighborhood's decarbonization

Fig. 10.8 Capa generates a new space for learning and production by adaptive reuse of an industrial existing structure

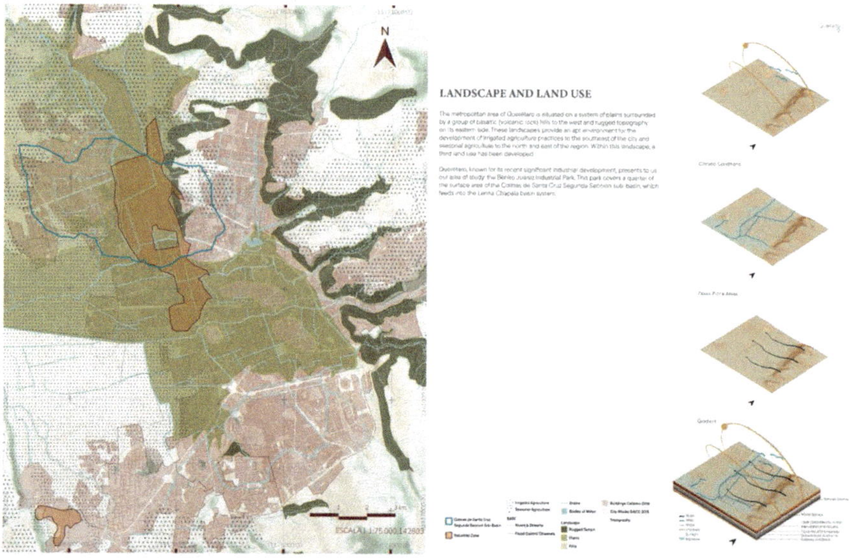

Fig. 10.9 Soil existing conditions

10.3.4 Soil

Soil and land are vital resources that sustain life in multiple dimensions. Proper management is crucial for environmental, economic, and social sustainability, especially in the current climate challenges. Soil is essential for its food production capacity, climate regulation, and water cycle because it is a habitat of organisms besides its nutrient recycling. In this context, we question how architecture and landscape architecture, in the context of the climate crisis, contribute to land regeneration and generate more resilient environments.

The current studio focuses on an industrial area in Querétaro for this topic (Fig. 10.9). The first industrial park in the region has visible effects of the industrial transition, leaving opportunities behind for a more resilient development. Due to polluted soil resulting from the transformation industry, it is an excellent opportunity to explore the potential regeneration of the site, contributing to the integration of a segregated and degraded area. The main challenges for this project will be soil degradation caused by industrial pollution and erosion, as well as the challenges related to the new programs to generate safer, healthier urban environments.

10.4 Discussion

Throughout this text, we have proposed exploring desirable futures in the face of the city's current challenges and those related to climate injustice. As a final reflection, the definition of climate justice design themes and the methodology implemented form a particular framework for dry urbanism.

While we understand the particularities of the context, academic speculations are excellent means of reflection, discussion, and training to create awareness of the complexity of future challenges. As Carracedo García-Villalba (2021a, p. 116) states, "...it is necessary to create awareness and expose future professionals in urban design, planning, and architecture to think critically and design cities in unpredictable and uncertain scenarios in light of the environmental vulnerability and the exponential consequences of rapid urbanization and climate change".

Furthermore, design studios are the backbone of architectural education and opportunities to encourage a holistic approach to complex problems such as climate change. As Sanderson and Stone (2021a, p. 2) outlined, design studio as the "backbone of architectural education" is a rich theoretical and practical discourse field. However, they consider that "current processes, theories, content, methods, and tools need to be questioned and critically examined." He argues that most design studios do not encourage a holistic approach to architectural education. He then lists the possible missed opportunities, such as an "awareness of socio-cultural and environmental issues, collaboration and teamwork, dialogic learning, sensitivity to and awareness of difference, critical discourse, innovative design, and technical competence."

This challenge requires holistic approaches with multidisciplinary teams and the possibilities of international collaboration. In these terms, "....the involvement of multidisciplinary teams formed by universities, researchers, experts, professionals, public agencies, community groups, and industries worldwide has allowed dealing with the complex task of integrating and bridging the gap between teaching, research, practice, and implementation" (Carracedo García-Villalba 2021a, p. 116). Within this context, international collaboration is an added value. We recognize that exchanging pedagogical practices and collaborative learning among different contexts holds the potential for intercultural knowledge and collective problem-solving. The convergence efforts serve as a platform to nurture resilient urban futures by sharing practices and learning toward the collective construction of knowledge. "Schools and faculties need to implement policies that promote teaching excellence, encourage a student-centered curriculum, support international dialogue and collaboration (especially in this time of uncertainty), and create global citizens (Sanderson and Stone 2021a, p. 2)."

Additionally, we recognize the value of implementing Research by Design (RbD) as the primary method. According to Hauberg (2011), research by design is a method that interconnects design and research, producing new knowledge about the world through the act of designing (Hauberg 2011). In these terms, the design process forms a pathway to create new knowledge through critical thinking, producing different outcomes and narratives that promote multidisciplinary interchange and dialogue.

The design studio 'Projects for the Future City' has developed nearly 100 academic projects in Queretaro, considering various scales with cultural, environmental, urban, and social perspectives. These projects aim to revitalize the environmental and urban setting, dignify spaces, and enhance the community's well-being. Even though we only show a selection of them, they result from *interdisciplinary approaches integrating climate science, urban ecology, landscape architecture, and urban planning* (Table 10.1).

Table 10.1 Inventory of projects, climate justice design teams, and tools

	Framework	Methodology	Theme	Date	Projects	Partnering universities	Multidisciplinary team consultancy	Manuals /tutorials/tools
Projects for the Future City	Climate justice design themes	Research by design	Water (hydric stress)	2023 (Spring)	Queretaro River	Universidad Autonoma de Querétaro	Hydrologist biologist, NGO, architect urbanist, landscape arch.	Crime Prevention through Environmental Design, LID, Guia Ilustrada Flora Valle de Queretaro
					Pueblito River	Universidad Autonoma de Querétaro	Hydrologist biologist, NGO, architect urbanist, landscape arch.	Crime Prevention through Environmental Design, Low Impact Develpment
			Urban Heat Island	2023 (Fall)	Citiies of Care	Universidad Central de Chile	Centro GEO, architect urbanist, landscape arch.	The Care Collective, Low Impact Development
					Via Verde	Universidad Central de Chile	Centro GEO, architect urbanist, landscape arch.	Heat Action Planning Guide, Tackling Extreme Heat in Santiago, LID
			Energy	2024 (Spring)	Punto	UNAM / Kent State University	Energy expert, architect urbanist, landscape arch.	Energy efficiency, District Energy Cities, eQUEST
					CAPA	UNAM / Kent State University	Energy expert, architect urbanist, landscape arch.	Energy efficiency, District Energy Cities, eQUEST
			Soil	2024 (Fall)	In process	UNAM	Geologist, soil experts, architect, urbanist, landscape arch.	Ecologies of prosperity, Suelo/Soil

Source: Authors.

These projects embody Querétaro's multifaceted narrative, in which resilience and innovation converge to shape a more equitable, sustainable urban future. Through a symbiotic interplay of academic research, community engagement, and nature-based solutions, they illuminate pathways toward a regenerative paradigm in which the city's dichotomous identity blossoms into a harmonious tapestry of human and ecological vitality.

Acknowledgments The authors recognized the valuable contribution of the teaching team: Viviana Margarita Barquero Díaz Barriga, Pedro Mendoza Hernández, Andrea Parga, Roberto Cevada, Miguel Anaya, and Daniela Cruz Naranjo.

References

Armas Sámano A, Ramírez Cabral D, Márquez Castro V (2024) Punto [undergraduate thesis, Tecnologico de Monterrey]. https://issuu.com/projectsforthefuturecity/docs/100624_book_digital_punto

Calvin K, Dasgupta D, Krinner G, Mukherji A, Thorne PW, Trisos C, Romero J, Aldunce P, Barrett K, Blanco G, Cheung WWL, Connors S, Denton F, Diongue-Niang A, Dodman D, Garschagen M, Geden O, Hayward B, Jones C, … Péan C (2023) IPCC, 2023: climate change 2023: synthesis report. Contribution of working groups I, II, and III to the sixth assessment report of the intergovernmental panel on climate change [Core writing team, Lee H, Romero J (eds)]. IPCC, Geneva, Switzerland. (first). Intergovernmental panel on climate change (IPCC). https://doi.org/10.59327/IPCC/AR6-9789291691647

Carracedo García-Villalba O (2021a) Pedagogy + resilience designing resilience in Asia international research program. In: Sanderson L, Stone S (eds) Emerging practices in architectural pedagogy: accommodating an uncertain future. Routledge/Taylor & Francis Group, London/New York, p 116

Carracedo García-Villalba O (2021b) Pedagogy + resilience designing resilience in Asia international research program. In: Sanderson L, Stone S (eds) Emerging practices in architectural pedagogy: accommodating an uncertain future. Routledge/Taylor & Francis Group, London/New York, p 121

Colunga ML, Cambrón-Sandoval VH, Suzán-Azpiri H, Guevara-Escobar A, Luna-Soria H (2015) The role of urban vegetation in temperature and heat island effects in Querétaro City, Mexico. *Atmósfera* 28(*3*):205–218, ISSN 0187–6236. https://doi.org/10.20937/ATM.2015.28.03.05

González-Sosa E, Ramos-Salinas NM, Mastachi-Loza CA, Becerril-Piña R (2013) Effect of water resources in the Queretaro River: climate analysis and other changes. In: Singh BR (ed) Climate change—realities, impacts over ice cap, sea level and risks. InTech Open. https://doi.org/10.5772/55259

Hauberg J (2011) Research by design – a research strategy. AE Revista Lusófona de Architectura e Educacao 5:46–56. http://revistas.ulusofona.pt/index.php/revlae/issue/current

Huerta D, Suarez A (2019) Cities of care [undergraduate thesis, Tecnologico de Monterrey]. https://issuu.com/projectsforthefuturecity/docs/cities_of_care

IMPLAN (2020) Plan Querétaro 2050 Plan estratégico del municipio de Querétaro a largo plazo. Recuperado en. https://implanqueretaro.gob.mx/v2/ot/p2050-inicio

Macías R, Bolaños DVV (2023) Trabajo final de pregrado, Tecnológico de Monterrey.. Accessed 10 Aug 2024. https://issuu.com/projectsforthefuturecity/docs/vi_a_verde_final

Olivera N, Amieva J (2021) Piezas de Tiempo [undergraduate thesis, Tecnologico de Monterrey]. https://issuu.com/thecityproject/docs/piezas_de_tiempo_2

Sanderson L, Stone S (eds) (2021a) Emerging practices in architectural pedagogy: accommodating an uncertain future, Routledge focus on design pedagody. Routledge/Taylor & Francis Group, London/New York, p 2
Sanderson L, Stone S (eds) (2021b) Emerging practices in architectural pedagogy: accommodating an uncertain future. Routledge, p 95
Tumini I, Higueras García E, Baereswyl Rada S (2016) Urban microclimate and thermal comfort modeling: strategies for urban renovation. Int J Sustain Build Technol Urban Dev 7(1):22–37. https://doi.org/10.1080/2093761X.2016.1152204

Chapter 11
From Oasis to Paradise: An Introduction to the Multiscale and Multifaceted Adaptation Aspects of Iranian Architecture and Urbanism in Dry Cities

Solmaz Hosseinioon

Abstract The world is suffering from increasing effects caused by climate change as well as disasters such as extreme heat, droughts and floods. There is a rising need for new approaches, and methodologies to (re)design the urban and rural environments for resilient, sustainable, and livable built environments. The methodologies and approaches used in Iranian architecture and urban design traditions relearn and reuse their adaptation methods to achieve maximum adaptability and comfort in extreme heat and scarce water resources though natural and energy sufficient ways. This chapter will give a contemporary insight into designing our cities and landscapes for dry conditions inspirated by examples from millennia and centuries ago.

Iran is a country with different climatic regions, each with ancient architectural and urban design and planning traditions for climate adaptation, creating thermal and environmental comfort for their residents. There are brilliant examples of adaptive methods from the building and housing scale to neighborhoods, landscapes, and city design. They have helped Iranian citizens outlive the harsh climatic conditions and lack of water throughout history.

The architectural methods developed in hot arid zones of Iran applied consistent building techniques which promoted passive climatic conditioning of the environment providing shelter and comfort from their natural environment. However, the Iranian cities are losing their building traditions and methods due to homogenizing urban regulations applied throughout the country whereas the neighboring countries are acquiring and claiming these methods as their own.

This chapter introduces examples of how Iranian houses, neighborhoods, cities, and villages have coped with dry, hot conditions, with a special emphasis on using natural elements: water, wind, sun and soil for maximum comfort and the best utilization of scarce resources. Examples include creating paradises (Pardis means heaven in Persian) in courtyards and cities (Persian gardens in the cities) by using water and vegetation inside the buildings creating shade and light rhythms in public

S. Hosseinioon (✉)
Independent Researcher, Melbourne, Australia
e-mail: solmaz.hosseinioon@rmit.edu.au

R. Roggema (ed.), *Dry Urbanism*, Contemporary Urban Design Thinking,
https://doi.org/10.1007/978-3-031-77347-1_11

spaces to cool the cities. It will also include a multiscale analysis of the methods from building scale to neighborhoods and city, underground, ground level, above the ground. The lessons learned can be applied to adapting to today's pressing issues we are facing towards adaptability, sustainability and resilience.

Keywords Iranian · Persian · Architecture · Hot-arid climate · Adaptation · Climatic design · Resilience · Urbanism

11.1 Introduction

Iran is a historic country with versatile climate regions. It has an ancient culture with resilient people who have learned to live with harsh climatic conditions of the country, from hot humid to hot arid zones to cold climatic zones spread from south to north. The historical architecture and urban from in these cities have been adapted to extreme hot and dry climatic conditions with scarce water resources and heat stress without using any mechanical added devices to ensure the maximum comfort and adaptation with the local conditions. This chapter will study the methodologies and approaches in Iranian architecture and urban design traditions to relearn and reuse the methods to achieve maximum adaptability and comfort in extreme heat and scarce water resources though natural and energy sufficient ways. This chapter will give a contemporary insight into designing our cities and landscapes for dry conditions inspired by historic examples.

11.1.1 Climatic Conditions of Hot Arid Regions in Iran

Iran's climatic diversity includes three main types:

1. Hot, dry and semi-dry climate: large parts of internal lands and southern border of Iran have this climate.
2. Mountainous climate, which itself is subdivided into two categories of cold and moderate mountainous climate.
3. Caspian climate: narrow and small area between Caspian Sea and Alborz Mountain Belt with 600–2000 mm annual rain.

A large part of the Iran central plateau climate is hot, arid, and semi-arid, especially in the central and eastern regions except the northern coastal areas and parts of western Iran (Fig. 11.1). Two Iranian deserts (the Dasht-e Kavir and Kavir-e Lout) occupy one-seventh of the total area of Iran, which is barren with very little annual precipitation (Tavassoli 2002). Many historical cities are in this hot and arid region (Keshtkaran 2011). The hot arid climate in Iran is mainly in the eastern and central parts of Iran, the main climatic characteristics include:

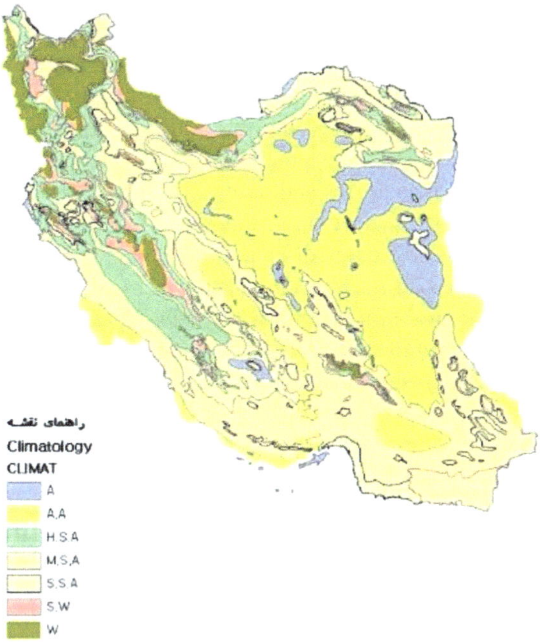

Fig. 11.1 Iran's climatic map. (climatology.ir)

1. Hot dry weather in summer and cold dry weather in winter
2. Very low rain
3. Low humidity
4. Meagre plant coverage
5. Big temperature difference between day and night
6. Dust storm around desert regions.

Water is generally scarce in most of the country except the borders of Caspian Sea in north of Iran and parts of coastal areas of Persian Gulf. The average annual rainfall of the country is about 240 mm with maximum amounts in the Caspian Sea plains, Alborz, and Zagros slopes with more than 1800 and 480 mm.

11.1.2 Historical Climatic Approaches for Adaptation in Architecture and Planning

The Iranian architects and builders have found smart solutions to deal with hot summers and cold winters of the desert. In some cases, dust storms and dry and hot winds have led people to creative designs for adaptation (Tavassoli 2002). Due to lack of humidity and water, life conditions are very hard in central parts if Iranian plateau. The historical buildings and cities in Iran were built according to adaptation with climatic conditions, making the best use of natural elements for environmental comfort. Adaptation to climatic conditions is achieved by incorporating natural elements: water, sun, wind, soil and local materials.

The construction methods used for adapting to this climate include using local materials such as clay and to build in creative ways, in accordance with climatic conditions. The vernacular Persian architecture uses form and architectural elements such as arches and domes, location and building orientation; and adopts the characteristics of outdoor, indoor, and semi outdoor spaces. The main principles include, introversion, orientation, going into the ground, house design according to the seasons, use of vegetation and water to create microclimates, using proper materials for insulation. Use of water networks is one of the key elements in shaping the morphology of the city (ZadehEsmail and Torabilangari 2013). This chapter introduces architectural styles and methods for building and design at the building and urban scales.

11.2 Architectural Elements and Methods for Building

In historical Persian architecture, the provision of a buildings indoor thermal comfort conditions was achieved by passive cooling systems, considering the building form, orientation, and selection of local materials. Architectural designs were developed, which incorporate passive environmental solutions related to the climate conditions to achieve thermal comfort with least energy consumption.

The harsh climate has made It necessary for Iranian people to search for effective solutions over thousands of years to live comfortably in a climate with the scorching sun, excessive heat, high temperatures during the day, and low temperatures at night, hot summers and cold winters, the arid weather and very little rainfall, dehydration, and frequent sand storms and they have managed to have great success in reducing the annoying aspects of the hot arid environment and create favorable conditions (Keshtkaran 2011). The main strategies to cope with this hot arid climate were the use of water and vegetation and making use of underground spaces (Bahrpeima and Sargazi 2022). Traditional vernacular houses in hot dry regions of Iran work as a complex passive cooling system that modifies the internal climatic conditions (Foruzanmehr 2014).

The architectural elements of the houses in these regions have been formed according to the climatic factors, so that construction of courtyard houses of these regions with the indicators such as thick walls of porches, basements, wind catchers, arches and domes show that architects had an explicit concept of environmental conditions. Traditional Iranian architecture in the past exhibited the following characteristics:

1. Introvert orientation.
2. Incorporating humanism and epistemology.
3. Use of symmetry and asymmetry.
4. Attention to nature and geographical climate.

5. Commitment and respect for maintaining connections with the urban context.
6. Use of compatible and cohesive local materials.
7. Use of natural light and color

11.2.1 Building Materials

In hot-arid regions of Iran, local materials are used such as mud, mud- brick and brick because of the climatic conditions. Unlike wood, they are available as the main local material. In Iranian architecture, this construction approach is used to reduce the heat and to make some changes on the appearance of buildings. Also, it disguises the lack of smoothness of the walls. The best indoor heat balance can be gained by using heavy construction materials in the living areas, which are used during the day, and light construction materials in the bedrooms where people sleep at night. The use of the building materials with high thermal capacity and light colors and suitable orientations are among the main design strategies for adapting with hot arid climate (Salehpoor et al. 2013).

Adobe and raw mud have high thermal capacity, they are heat resistant in hot summer days and provide insulation keeping internal heat during the nights and cold-dry seasons. It takes a long time for the heat outdoors to pass through the walls and get into the indoor spaces of the houses. Adobe can delay the heat transfer from outside to inside for about 8 h, which means the heat that is accumulated in the walls during the day will warm the house at night when the outside temperature decreases in arid regions where the temperature drops at night.

11.2.2 Courtyards

Courtyards are among the central elements in the architectural design of the buildings throughout the history in this region used from pre-historic times in central Iran and in other parts of the world. The courtyards are one of the most common elements of buildings in Iran and have had a significant role in Persian architecture (Vakilinezhad et al. 2013), as well as in the Middle East, and most ancient civilizations. Houses with central yards are the most beneficial form for decreasing the exposure to harsh weather conditions, especially in hot, arid regions and where the sand storms coming from desert. The rooms that open to the yard are usually protected against the extreme heat of the summer, the cold of the winter, and from wind, storms, and sand in desert regions (Meamarian 1999; Keshtkaran 2011).

This internal open space of the courtyard in the houses is substantial in the physical and spiritual life of the people have had a great role on thermal comfort of the buildings. This introvert space is intricately linked with the typologies, geometric design, proportions, shading, fenestration, functional arrangements, shading, surfaces and materials, direct contact with nature, lifestyles, and rhythms in Iranians' lives for centuries and has acted as a pivotal role in social and psychological wellbeing of the residents.

Courtyards have deeply affected the form, and the local functional patterns, and maximum use of empty and solid spaces (Soflaei and Shokouhian 2008; Piryaei et al. 2022). The courtyard shape not only creates an ecological space for breathing the building but facilitates the connections between the interior spaces and minimizes the internal movements (Soflaei 2012).

The courtyards and yards have lot of potentials such as providing light, shade, natural ventilation, reducing energy consumption, and thermal comfort (Soltanzade and Sadatzad 2013, 2) They are microclimate modifiers, which help improve the thermal comfort in the enclosed as well as the attached built volume (Ebadi et al. 2014). This means that water and vegetation in addition to shading created by building facades in the courtyard play a significant role in creating climatic comfort and humidity in Persian traditional houses. They provide thermal and climatic comfort, as small oases inside every house which include presence of water in the form of ponds call Hoze, and in bigger houses streams, called Jooy. They help with cooling the space via evaporation as well as creating a pleasant sound and sight for the residents.

For example, in central courtyards of Kashan houses (the Borujerdi and Tabatabaeiha houses) a considerable percentage of the courtyard is dedicated to a water pond (Hoze) and greenery (Fig. 11.2 Borujerdi house). This means that the combination of water, greenery and shading created by the surfaces surrounding the

Fig. 11.2 Wind-catchers in Iran's central city of Yazd. (www.japantimes.co.jp)

Fig. 11.3 Borujerdi house. (https://bashgahfarhangian.com)

courtyards, have a great role in creating humidity and environmental comfort in the traditional houses of Kashan (Jafarian and Monsefi Parapari 2021).

This space has many functions including providing air flow, light, connection between spaces in addition to its own special functions (Zareei and Mirdehghan 2016). These climatic solutions, are achieved by a combination of water and greenery and in 45 °C average heat, creates cool environment which reduces the energy consumption (Baghai et al. 2014; Bahrpeima and Sargazi 2022).

The living rooms surrounding the courtyards usually have a high ceiling and are connected to the bedroom on the first floor, which help the hot air rise and spread on the first floor. On sunny days, the outdoor and indoor curtains are lowered, while, after sunset, they are raised. Hence, in summer, the process will be reversed (Fig. 11.3).

11.2.3 Orientation

One of the main factors affecting the design of houses and buildings include orientation, climate and form. The orientation of the house in a hot, dry climate should be designed in a way to protect the building from the hot sunshine on summer afternoons (Keshtkaran 2011).

The direction and location of the buildings in the lots are chosen according to climatic situations with an emphasis on the direction and angle of sunlight in summer and winter. The architectural styles of the plans aim to make buildings dense and complex, providing maximum protection from the cold weather at night and in winter and hot weather in the summer, casting shadows on the house walls and across the streets (Tavassoli 2002; Keshtkaran 2011).

In Iranian traditional architecture, natural light plays a fundamental and significant role. In hot-arid climates, the dense and intricate building placements are aimed to reduce the high sun heat on exposed facades. The building layouts and plans are built in a compact and dense way in this climate to reduce the maximum contact and temperature exchange of external walls with outside environment causing maximum shade creation on the surfaces.

One of these crucial principles in traditional architecture, which has been neglected today, is proper building orientation. This refers to the optimal direction of a building.

Southern light walls, positioned to maximize solar radiation during winter and short days of the year, receive direct sunlight. In summer, especially in hot and dry regions, they receive less direct sunlight. Since optimal sun orientation is rarely aligned with wind direction, buildings should be positioned to protect against undesirable winds throughout the year.

Moreover, during hot seasons, natural ventilation and temperature reduction can be achieved using favorable breezes. One influential factor in choosing this orientation has been the proximity of cities to one or several mountains.

Natural ventilation occurs by wind coming through open north and south windows during the evenings and nights in summer, and the cool air is saved in the house for the morning hours. In the morning, while all the windows are covered by the curtains to prevent light from coming in, the window of the main bedroom is open and acts as a chimney to remove the hot stream of air, and the cool air of the north direction and deep-ground yard replace the hot air (Gharehgolchian 2010; Keshtkaran 2011).

Most of the openings and fenestrations are open to the inner courtyards which reduces the contact of the building with the outside harsh environment, which is also defined with an emphasis on form, direction also called" Roan".

The subject of Roan (orientation) is highlighted as one of the features of traditional Iranian architecture, which refers to the orientation of buildings considering climatic conditions such as sunlight and wind exposure. Iranian architecture is categorized into three orientations or "roans":

1. Roan-e Raste (Straight Roan)
2. Roan-e Esfahani (Isfahani Roan)
3. Roan-e Kermani (Kermani Roan)

These three orientations have been considered in the architecture of many historical cities in Iran. According to Dr. Pirnia (1993, 1999) traditional architects utilized placing a Regular Hexagon in the Roan-e Raste and Roan-e Esfahani orientations, the rectangle within the hexagon deflects 30° from south to west or east. The Roan-e Raste is described as northeast-southwest and approximately aligns with the qibla (direction of Mecca) in architectural design, which has been considered in the cities of Tehran, Tabriz, Yazd, and Jahrom. Pirnia (1993, 1999) considered any orientation other than Roan-e Raste highly unsuitable for the city of Yazd.

Although today, in new areas of Yazd, constructing houses in an inappropriate east-west direction has led to excessive heat in summer and severe cold in winter. If the building is to be constructed in one direction, 25° east is the best direction for the main body of the building, but this can be changed to as much as 35° southeast.

11.2.4 Light and Shade

Traditional architecture has a variety of solutions of light absorption according to the type of use and the climate of the area. In this way, buildings in cold climates benefit from the maximum sunlight. Buildings in such climates are placed in such a way that they receive the most sun exposure. Elements of light regulation and control in traditional architecture play a role as intelligent measures in controlling light within buildings which in turn control heat and glare inside the buildings. They include numerous creative elements such as roozan (small windows), shabak (screens), lattice doors and windows, jame-khaneh (glass domes), hornu (skylights), ersi (latticed windows), roshandan (lanterns), khoon (decorative tiles), goljam (rosettes), and palakaneh (small glazed openings) not only enhance beauty but also serve functional purposes.

11.2.4.1 Shabak

Shabak is a type of built lattices acting as curtains or screens for protection against external elements besides doors and windows in buildings, they control intense sunlight, variable weather conditions, wind, rain, storms, and specific national and religious beliefs. The function of shabaks was to obstruct the view from outside to inside the building, although it allowed visibility from inside to outside space.

11.2.4.2 Rozan

In traditional Iranian architecture, roozans are present in the calyxes of domes, warm rooms, clusters, and markets as glazed rings in the form of domes or projections. In winter, glass round cups like the bottom of a cup are placed among the glazed rings and in summer, one or all of them are removed. Rozan can be a small window above the door. This window is used to provide illumination and sometimes to allow free airflow in enclosed spaces. Roozans refer to holes embedded in the calyx or neck of domes. The calyx is the head or peak of the dome. These roozans have also been used in inward-looking structures alongside the central structure. Inside the buildings, they used roozans or wooden and plaster windows, covered them with ceramic or tiled screens on the outside. These screens reduced the intensity of light and directed a softer light inward through their patterned designs.

11.2.4.3 Lattice Doors and Windows

In traditional architecture, lattice windows are prominently seen in areas exposed to intense sunlight. These lattice doors and windows contribute to creating balanced illumination within the building. Additionally, they facilitate airflow and provide views outside the house. In this case, these kind of roozans are also called roshandan. Roshandans have glass and some have polygonal backgrounds. For example, we can refer to the eight-paradise pool lanterns in Isfahan.

11.2.4.4 Orsi

Orsis are latticed windows that open upward instead of swinging on their hinges. Orsis are seen in vestibules and porches of cold-climate buildings and alcove chambers. Orsi screens, like wooden windows and roozans, serve a functional purpose.

11.2.4.5 Hornu

The skylights above the ceiling, hornus, are called light catchers. They can be seen in the ceilings of old markets. They look like holes open to let in lighting and ventilation and air passage because it was not possible to install windows in walls in markets and public buildings.

11.2.4.6 Khoon

Khoon has a decorative aspect in the building. Khoon refers to decorative tile-like images made on the front of the building. In this technique, they use the thickness of the brick to create various decorative nodes on the front facade. Valuable parts of the building such as the entrance head, internal courtyard facades, especially the main ivan parts, and in some cases the exterior walls of the house are decorated with khoon.

11.2.4.7 Karbandi and Moqarnas

Karbandi and Moqarnas, are not only structural aesthetically amazing elements, but they also function well in helping to spread light in space. When lighting is provided through the roof, the ceiling Karbandis divert light in different directions. As a result, more light is uniformly and non-concentrically emitted in space.

11.2.4.8 Hashti

Hashtis are spaces that are circular or polygonal and are embedded in the entrance of the building (Figs. 11.4, 11.5 and 11.6) Often at the top of these spaces, there have been light catchers that have directed soft focused light in different hours of the day (Eskandari 2011).

Fig. 11.4 Tabatabaeiha. (irangard.com)

Fig. 11.5 Hashti: Space hierarchy: entrance (1), corridor (2) and vestibule (3). (Eskandari 2011)

Fig. 11.6 View of entrance to Hashti Tabatabaeiha house https://iranmonument.com/%D9%87%D8%B4%D8%AA%DB%8C/

11.2.5 Use of Underground Spaces

In critical conditions, buildings are built underground or in the heart of the hills so they will be less affected by the adverse weather conditions (Meamarian 1999; Moradi 2005; Keshtkaran 2011). These materials can be recycled and thus are sustainable.

The basements are extensively used both as a place to keep food, access the Qanat waterways if possible. If the house is located to directly access clean cool water from the inside of the house and in some case, it is a sitting hall underneath the building which stays cool even in hottest days of summer. It is the coolest part of the house, and it is even made cooler since in some cases, a pond or water way is running in the middle of the hall or the wind from the wind catcher (Badgir) passes through it. The Sardab is always cool, so it is enjoyable to spend time in hot summer days. It is so cool that in occasions, it can sometimes even feel uncomfortable.

11.2.5.1 Godalbaghche (Pit-Garden/Pit-yard)

In some traditional houses, the yards are dug deep in the ground and called Godalbaghche. By building Godalbaghche in the ground, the humidity and shade are increased (Moradi 2005, Keshtkaran 2011). They are usually located in the middle of the courtyard and is sunken in the ground about one level, such as in Kashan, Yazd and Naeen. There is usually streaming water which fills the central pond. There are rooms around it and there are fruit trees, and its climate is usually cooler and more humid than the ground level courtyards. A great example of this is the BozorgAgha School in Kashan. The Godalbaghches are built as an opening below the ground level, the walls of the rooms surround all four parts of the house, so it sits in the shadows on hot summer days. The main reason for creating *Godalbaghche* is increasing shade and helps to keep the air cool air in the yard during the day. In addition, at night when the weather is cool, the deep pit-yard becomes cool and stays cool for several hours, even though the sun is shining almost directly from above. Sometimes the depth of the yard is as much as 3–4 meters (Meamarian 1999; Keshtkaran 2011). In some cases, they also facilitate access to the subterranean canals, Qanats. They are used to water the plants and gardens in the yard. The depth of the yards was more than the normal to have access to Qanat water or subterranean canal of water, which was passing underground of yard, to water garden and below ground courtyard. These yards were called Padiav. By having a cover full of plants and trees and naturally because of evaporation they are acting as a cool and fresh air generator for the upper yard spaces (A'zami et al. 2005).

11.2.5.2 Sardab

Sardab is an underground space usual built underneath the summer hall and its ceiling is 1 m to 50 cm higher than the ground level. The ceiling of the Sardab is usually 1 m higher than the yard level so that it has openings (as windows or lattices) to the yard which let in the cooler air on the ground into the Sardab which in turn becomes cooler by passing over the pond or the stream derived from the Qanat, one of the best examples is Abbasian House in Kashan. It has a pond and an underground stream or sometimes access to Qanat water in the center. The Abbasian House in Kashan (Fig. 11.7) is a prominent example (ZadehEsmaeel and TorabiLangari 2013).

11.2.6 Temporal Use of Building/Houses Elements

Seasonal usage of spaces, concentration and attention to courtyard and making suitable use of roof are the highly functional ways of planning in accordance with hot-arid regions (A'zami et al. 2005). Nature is an inseparable part of people's lives. Summer and winter areas in Iranian houses can be a seen as a replication of Nomad's movements across Iranian Plateau. This complex modifying system has various components or sub-systems of which one is the existence of separate seasonal rooms in the house (Foruzanmehr 2014).

The residents move to different parts of the house in each season to adapt and make best use of sun orientation, and warmth as well as making best use of dominant wind (or prevent its harsh effects). So, houses are divided into two parts,

Fig. 11.7 Tabatabaeiha house Hashti https://lotusclock.com/traditional-iranian-architecture/

Fig. 11.8 Abbasian (Abbasi) House, Kashan, https://en.wikipedia.org/wiki/Abb%C4%81si_House

a summer area, and a winter area (Fig. 11.8). This feature can be seen in the houses in most cities of this region. In most houses in Yazd, the main part of the house faces northeast. In the summer, the main part of the house is the summer area, which is in the shadows most of the time. Across the yard from the summer area is the winter area, which provides access to the warmth of the sunlight during the winter. The owner builds the two other parts depending on her or his desires (Keshtkaran 2011).

To prevent the extra heat of summer, especially during the very hot afternoon period, the summer section (the part in which families spend most time) is built with its back to Qeble (Southwards Mecca), which is called "Nesar", meaning south direction in the building in local dialect. The other side, which is in front of Qeble, is perfect for the winter sunshine.

11.2.7 Proportions

One of the key principles of Persian architecture is proportion. It has close relationship with climatic conditions. "Peymoon" in Iranian architecture served as a method to facilitate work and standardize all measurements in construction. It diversified architectural designs, ensuring that no structure appeared derivative. There are three types of "Peymoon" or scales in Iranian architecture: small "Peymoon," large "Peymoon," and minor "Peymoon." "Peymoon" not only influences the dimensions of foundations, columns, room widths, and corridors on plans but also determines the proportions of doors, windows, and the configuration of portals, iwans, domes, and house vaults. Its impact is evident in enabling Iranian architects to confidently design, calculate, and execute projects within a unit without compromising stability.

"Peymoon" is a collection of measurements, quantitative ratios, or qualitative assessments that have been implemented in the architecture of various structures in traditional Iranian architecture since ancient times. This framework, akin to a universal pattern in all types of structures ranging from palaces, fortresses, citadels, and pavilions to reservoirs, bathhouses, bridges, markets, and houses, has had significant practical utility. All architects and master builders throughout the cultural expanse of Iran regarded it as an indispensable guideline, seeing it as irreplaceable and approaching it with responsible attention. The right proportions of the courtyard created proper shade and create tangible thermal comfort and reduces the need for active cooling in summer (Zeynalian and Okhovat 2017).

According to Pirnia (1993,1999) the best proportion in a courtyard is one to three (Tayari & Nikpour 2022). A crucial aspect of "Peymoon" was its oral and face-to-face transmission from master to disciple over a millennia-old process, reaching the pinnacle of precision and efficiency. Therefore, apart from texts such as the treatise "Taagh" by Ghiyath al-Din Jamshid Kashani, which indirectly touches upon some aspects of "Peymoon," a specific document detailing these proportions and measurements is not readily available. The loss of this tradition of mentorship, coinciding with the decline of traditional Iranian architectural methods and their replacement by modern civil engineering techniques, has placed this significant oral tradition, which alone constitutes a substantial and valuable part of Iran's intangible cultural heritage, at risk of being lost in the throes of destruction.

11.2.8 Badgir/Windcatcher

The windcatcher is an innovative Iranian method for creating a cool space inside hot desert homes. This air conditioning device has made living spaces in Iran bearable for many years. Over the millennia and centuries, as old as Iranian civilization itself, buildings have been constructed with attention to the climate and environmental conditions. Sunlight, wind, humidity, cold, heat, and generally weather and geographical conditions have had a direct impact on traditional Iranian architecture in various regions.

The most prominent method of natural ventilation in buildings is the windcatcher. Windcatchers have different shapes designed and implemented in many central and southern cities of Iran based on the dominant wind speed and direction. The ancient and various names of windcatchers, such as Vatfar, Badhenj, Bathan, Khishud, Khishkhan, Khishur, Masoureh, and Havaakap, indicate that this phenomenon is very ancient. In 1969 AD, the famous archaeologist, Masuda, found two hills in his way from Shahrud to Gorgan and named them the Flint Hills. These two hills, which are called the eastern and western hills, are located in 8 km north Shahrud and near the southern slopes of Alborz. Masuda dug two hills during the years 1971 and 1973 AD and estimated the approximate date, 7800 years before now for it trough carbon-14 test (Kiani 2004). In plans, provided by Masuda from the explorations, he introduced the locations-of residential plans as the fireplace. In his article, he noted that the existence of ashes in some of these places is because they are fireplaces (Roaf 1982). The windcatchers working system is based on drawing the flowing air from outside into the house. Using water-filled basins placed inside them, they cool and lighten the air before directing it into the house.

11.2.8.1 Types of Badgir

Meamarian (1999) claims that in addition to their important functional role for passive cooling of the buildings, some Badgirs have decorative and symbolic values too. They can indicate the landlord's financial status too. There are sevral types of Badgir:

- One sided (Ardakani)
- Two sided (Kermani)
- Three sided
- Four sided (Yazdi)
- Chopoghi (pipe shaped-Sirjan)
- Eight sided (Dolatabad garden Yazd-33 m high)
- Three story four sided Badgir in Taft, Garden Sadri
- Two story and four sided Badgir in Garden Amir Tabas
- Two story and four sided Aghazade House in Abarkooh
- Two story Round Badgir in Chehelsotun Palace in SarhangAbad

11.2.8.2 The Method of Building a Windcatcher

Local architects construct the windcatcher from the rooftop of the house, starting from a point overlooking a small room dedicated to the windcatcher. Using clay or bricks, they build the rectangular-section shaft of the windcatcher until it reaches a certain height. Then, they place four walls on top of this shaft, using two pieces of wood in a crisscross "×" pattern, ensuring that each end of the wood rests in a corner of the rectangular section.

Natural elements of the cities in hot arid climates such as Yazd include harsh, high-speed winds and sometimes sandstorms. That's why architects must consider favorable and unfavorable winds. They use the favorable winds in hot seasons with the help of the Badgir. One of these favorable winds is known as the Isfahani wind, which moves in the northwest direction. The Badgir is an architectural element that has a climate function (Mahmoodi 2009).

11.2.8.3 Function of the Badgir

Wind-catcher acts by sucking and pulling. The basis of action is that wind blowing is used to suck the cold air to the inside of building and the reaction of it is used for sending out the hot and pollutant air from inside the building. Once the wind encounters walls of internal wings of wind-catcher inevitably it descends and enters the building space, on the other hand the holes or vents of wind-catcher on the opposite side of wind blowing to the sucking and give the hot and polluted air of building to the wind (Ghobadian 1998; Ghobadian and Mahdavi 1993).

A typical Badgir includes these parts: 1- the main body 2- the opening and 3- the blades (each of which different parts). Since the framework of the Badgir is built higher than the height of the building itself and the opening which sucks the air in, reduces its strength against lateral powers, the importance of resistance of the elements mentioned above is obviously significant. For example, a Badgir in Yazd, has a wooden ram is placed horizontally in bricks, and this obviously increases its resistance against lateral forces (Meamarian 1999; Keshtkaran 2011).

Badgir or as translated as windcatcher/wind tower in English, is an Iranian creation for creating coolness inside the houses in hot arid deserts. Badgirs are an ancient air-conditioning system,that has made life pleasant for Iranians for centuries. The technology has travelled through the Silk Road as well as by lands overtaken through wars and expansion of empires throughout history.

Badgirs are usually small towers with octagonal or symmetric polygons plans. The main function of Badgir can be summarized in two parts: attracting the wind to bring cool airflows inside the Badgir and into the lower part of the building. The second task is to push the hot airflows out of the building (Darani et al. 2013). The directions of windcatchers which can be one sided, two sided or four sided.

Water had to be located under the space beneath the Badgir for cooling purposes. The presence of water was incorporated in 4 ways:

1. In the underground space, there is a pond (Hoze).
2. Putting water in earthen jars.
3. Connecting the Badgir tunnel with a damp channel which is connected to the main sitting room, this channel is connected to the central garden in the courtyard and when the garden is water the humidity of the garden cools the weather inside the Badgir channel.
4. Putting layers if woven basket, clumps of desert thorns or earthen jars in the entrance of the Badgir and spray it with water. They hold the humidity for some time and cooled the air which entered the wind channel (Shamlou 2007).

Badgirs are usually located in the back part of the summer section of the houses (Saljoughinejad and Sharifabad 2015) through which air flows into the house, and, in some houses, the air is also directed to the basement. Having a pond in the yard and some plants or a little garden creates evaporative cooling, and, with the help of the Badgir, the cool, humid air flows in the house, creating a pleasant environment (Fig. 11.9). On summer nights, when people usually sleep on the roof or in the yard, the Badgir gets the cold, night winds and spreads the cold air all over the house, and since the doors are closed, the house remains cool during the morning (Tavassoli 2002; Keshtkaran 2011) (Fig. 11.10).

The functions of Badgirs are primarily to receive the favorable wind for cooling and ventilation and to direct it to spaces such as the main room, basement, and water storage area. The opening of the Badgir, which faces the wind, has positive pressure applied, and the opening on the side opposite to the incoming wind has a pressure that is less than the ambient atmospheric pressure (the suction part).

The middle septum of the Badgir separates the suction part from the pressure part. The wind that enters hits the vertical septum in the middle of the Badgir and directs the air to the indoors space through channels that face the wind. It should be mentioned that a group of Badgirs cool the indoor space just by convection, while others cool the air via evaporation in addition to the convection. A pool filled with water is built under the Badgir channels, so the wind that enters through the channels passes over the surface of the water and cools the air by transferring some of its heat energy into the water to increase the evaporation of the water. The increase in humidity that results from the evaporation in arid regions is of great importance. The air that is used indoors absorbs heat from the environment, and the reduced pressure in the back channels causes the air to be rapidly removed, causing an inflow of air through the front openings to replace the used air inside (Mahmoodi 2009; Moradi 2005; Keshtkaran 2011).

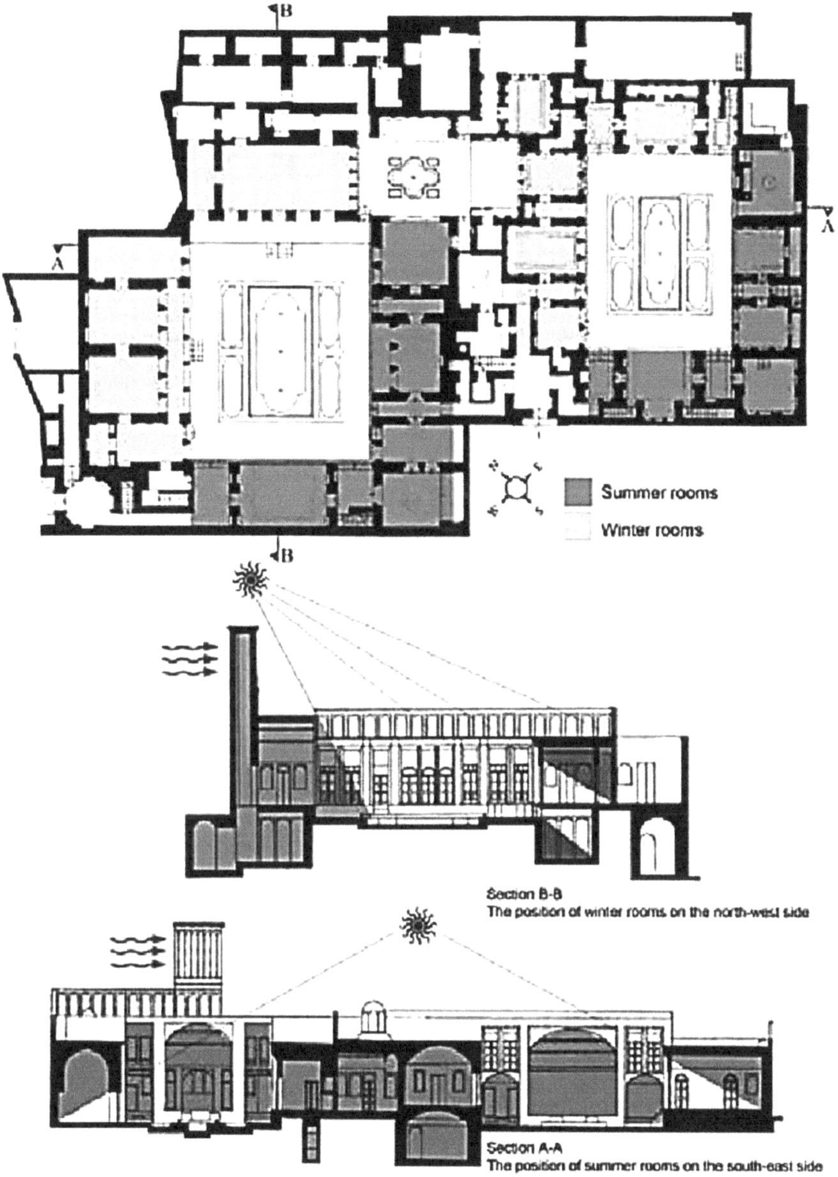

Fig. 11.9 Seasonal movement, summer and winter spaces. (Hosseini et al. 2015, Analyzed by author from http://www.naghsh-negar.ir)

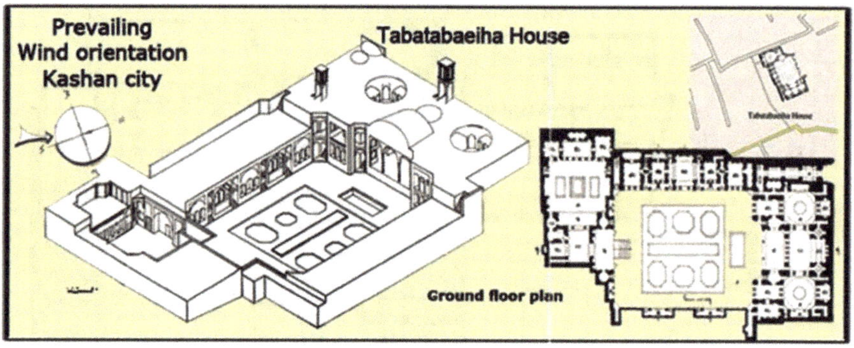

Fig. 11.10 Tabatabaeiha House design according to prevailing wind. (Saljoughinejad and Sharifabad 2015)

11.2.9 Volumetric Design of the Ceilings and Roofs and Details

The roof shape of the buildings in this region has several thermo-physical reasons. Due to having convex and unbalanced surface the impact angle of sunbeam on dome and arched roof is different from one point to another, and a part of it always remains in shade during morning and afternoon times. For this reason the curved shape is suitable for releasing and emitting sunbeams and waves during night and it helps to the night cooling. One reason for using high roofs in hot regions is that warm air rises, so high roofs can provide thermal comfort in the heat (Keshtkaran 2011).

The buildings' volumetric design in different scales from roof scape to details in interior design of the buildings are carefully designed to adapt the buildings to climatic situations. Arched or domes are the dominant form of roof using adobe and mud in this climate. The domed roofs means that they are constantly exposed to the flow of air caused by the wind. In addition, half of the dome is always in shade which makes it cool and encourages air flow due to temperature difference in different parts of the dome. It helps reducing the heat of the roof due to severe sun radiation. At night, it helps the roof to cool faster as well. In the domed or cylindrical roofs, the sun radiation is not the same on all sides. This is also useful in reducing the temperature under the roof (Moradi 2005; Keshtkaran 2011).

The roof shape of the buildings in this region has also thermo-physical reasons. Due to having a convex and unbalanced surface, the impact angle of sunbeam on dome and arched roof is different from one point to another, and a part of it always remains in shade during morning and afternoon times. For this reason, the curved shape is suitable for releasing and emitting sunbeams and waves during night and it helps to the night cooling. One reason for using high roofs in hot regions is that warm air rises, so high roofs can provide thermal comfort in the heat (Keshtkaran 2011). If the flat roofs are used in hot & dry regions, it is usually paved with square shape bricks called paved bricks. These bricks receive the most radiations of sun.

Fig. 11.11 Image 158104—UNESCO World Heritage Centre

It changes the sun radiation intensity and radiation angle (A'zami et al. 2005) (Fig. 11.11).

Locating clay bowls known as taghar on the roofs has a root in climatic solutions for hot arid conditions. They are called "Nahaz" (protrusion) and "Nakhir" (indentation) with an emphasis on bulk and space. Carvings in the thick wall which create shelves called: Taghche not only reduces the building load against side forces and ranesh break the sound waves. In traditional Persian architecture even the reduction or omitting of a small ornament such as shemshegiri can change the function and authenticity of a building.

11.2.10 Eyvan

Eyvan is a type of Veranda is a semi-closed space in front of the building opening to the yard and is usually part of summer part of the house. It creates a semi-closed space which not only creates a thermal shield but also creates shaded areas for use by the residents to enjoy the shade and wind and yet be protected from the sun. A useful and significant part, such as a porch as a half-open space, can be used in hot, arid regions in the summer. The climatic role of Eyvan, in protecting the building from heat includes creating shade on the building volume, providing ventilation and protection against heat (Zarei et al. 2017) the importance of the summer Eyvan is that it usually benefits from north/north-eastern sunlight and its depth is deigned according to the angle of the sun in summer. Eyvans are used in summer evenings when the sun passes below the front wall. They are usually elevated three or more steps higher than the yard, so it is slightly raised above the yard. The basement vent is under the steps of the summer area (Meamarian 1999; Keshtkaran 2011) (Figs. 11.12 and 11.13).

Fig. 11.12 Tag Kasra, https://www.atlasobscura.com/places/taq-kasra

Fig. 11.13 Typical Eyvan in a Persian house, https://foursquare.com/mlles1901848/list/sehenwurdigkeiten

11.2.11 Water

The element of water as a natural element and courtyards as one the key components of Persian architecture are combined to provide environmental comfort through evaporation and cooling in hot arid climates. The pond (hoze) and the underground channels that lead to the lowered northern yard. It streams the cool air to the living room with the help of the air that has been cooled by giving up its heat to evaporate water. In the summer, the cool air of the basement comes up through the vertical channels, and, in the winter, the fan and the heater in the basement provide warm air to the bedrooms through this channel.

Padiav is a surrounded yard which has a pond or stream located in the middle of it. Padiav means cleanliness and washing. This word has moved to Europe and transformed into Patio. It is also called Godalbaghche, meaning pit-garden.

The invention of Qanat/ Kariz has had a great role in the economic and social prosperity and creation of many jobs and development of cities and villages because the cities shaped in dry regions didn't have access to water like the ones located in mountainous areas of Iran. Qanat or Kariz was invented and used from Iron Age as a source of providing for drinking water and agriculture in the dry Iranian plateau. This ingenious irrigation system called Qanat or Kariz which was invented 5000 years ago, it is as ancient as history of Iran (the oldest Qanat called Jupare in Kerman is estimated to be built 5000 years ago (Karaji1010 AD). Its name was mentioned in Sargon's inscription as the cities' irrigation system.

The height of its use was in Achaemenes dynasty, especially in the era of Darius the first (486–521 BC) where the creation and use of Kariz was widespread and spread throughout the entire Achaemenes empire by the central government. Hence many countries around the world which were once a part of this great empire have used this invention ever since. In Ashkanian and Sassanid dynasties their use is so widespread that a special administration system was created to supervise their use and manage the water resources. Building Qanats continued after Islam and the technology was exported to the countries which were overtaken by Muslims. But after Mongol invasion, many of them were destroyed and in Safavid Dynasty, water scarcity became such a critical problem that the Qanats were revived again.

11.2.11.1 Abnama, Hoze (Pond)

The Persian architects put a pond in the middle of a house in hot-arid climates and created a pleasant microclimate and by putting the openings of the building, creates thermal and climatic comfort for them (Vakilinezhad et al. 2013; Ghobadian and Mahdavi 1993) by cooling the building through evaporation specially through evaporation through a pond and its fountain (Farrokhyar 2014; Zangeneh et al. 2014).

Use of ponds in the centres of Persian houses goes back thousands of years. Archeological excavations in Tappe Yahya in Kerman province (5000 BC) is one of the first examples of using water ponds in the center of the buildings (Encyclopaedia Iranica 1989). It was used in Mithraic temples which were used for purification and initiation ceremonies. Later they were used for washing and religious ceremonies in Iran after Islam as well. It represents heaven, paradise or Paradis in Persian and helps create a piece of heaven on earth for the household.

In addition to the religious reasons, one of the main factors for creating ponds and bodies and water in the houses is the geographical situation of Iran which many people, move to the mountains and cooler areas in summer, but when they can't do it, they tried to create cool places through passive methods.

The pond is built either in an enclosed area in the house, usually in a room dedicated to it called HozeKhane (house of the pond) which was used for sitting and entertaining in hot summer days. Sometimes this room is on ground level but in cases such as Yazd, they were located underground.

The second and more common use of the pond which has continued until contemporary times is when it is built in the middle of the courtyard.

Other than ceremonial and aesthetic use, the pond makes up for lack of humidity and makes the air pleasant (Soltanzadeh and Sadatzad2013). The pond shapes include circles, rectangles, hexagons, or polygons which affects the landscape design of the gardens in the courtyards (Nayebi 2002).

11.2.11.2 HozKhane (Pondhouse)

HozeKhane is an enclosed house on ground level or underground. In the first case, the light is provided from a skylight, and they usually cooled by Badgir and the combination of cooled weather and water evaporation from the pond and its fountain (Mahmoodi 2009; Zangeneh et al. 2014). The HozeKhane (Pondhouse) plan is usually either half-closed or closed.

1. Open plan: one side of HozKhane is open to the yard through an Eyvan, as a half-closed space, this shaded area is protected from the heat of the sun and is called by evaporation.
2. Closed plan: it has no connection to the yard and is surrounded by rooms.

Two horizontal channels that are perpendicular to each other, one from the storehouse in basement and the other from the lowered yard that is connected to the small basin, have a way to the living room, ending at a well with a depth of 3 m at their interchange point. The channel that leads to the pool is half filled with water, and the overflow goes into the well, always keeping it damp. So, the air gets humid at 24 °C and streams into the living room (Fig. 11.14).

In both cases the ceiling is dome shaped and the material is usually brick or plaster, in cities such as Kashan, it is called summer Talar and is separated from the building and are connected to the yard. But in cities such as Shiraz it is located between the rooms and are connected to the yard through corridors.

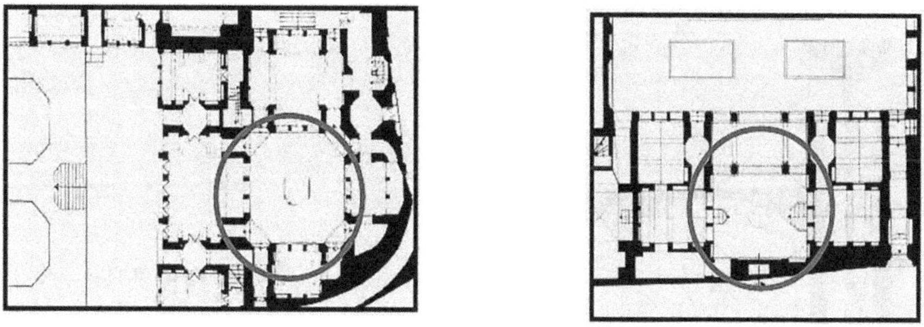

Fig. 11.14 Half-closed Hozekhane in Attar house and closed Hozekhane plan in AlYassin house

11.3 Neighborhood and City Design in Hot Arid Cities of Iran

The urban morphology and design of the Persian cities were shaped by the blue-infrastructure specifically the water distribution systems in Iran. The urban Morphology is dense with narrow lanes, a combination of buildings has led to decrease external surfaces and con serve the needed energy in internal parts.

The traditional cities in hot arid climates of Iran such as Kerman and Yazd have introvert and compact urban morphologies completely based on the environmental characteristics. The dense fabric, the narrow winding and shaded alley and streets and even the color of the building materials reflect this adaptation to climatic conditions. Winding streets with introvert squares form the central core of the traditional cities. Creating shade by tall walls, latticed windows and natural ventilation are among the main strategies for adapting to the climatic conditions (Asghari-Moghadam and Rajabi 2006).

The urban texture is condensed and compressed to each other in these regions. Houses have merged or combined walls and the border between them cannot be identified. The compression and combination of buildings has lead the external surface of each building to the least and as a result each home can conserve the needed energy inside of it for a long time. It is also an answer for the crowdedness. The narrow-serrated lanes, which sometime have high walls and are roofed by arches, cast a shadow on the surrounding houses and control the wind speed of the dessert (Kavir). The structure of city is planned in a way, which arteries are open in the direction of desirable winds and closed in the direction of undesirable winds and sandstorms (A'zami et al. 2005).

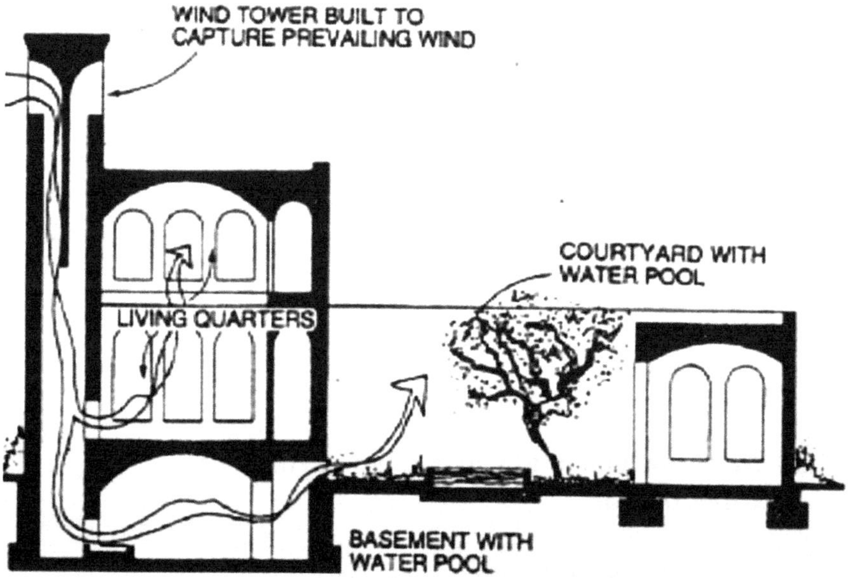

Fig. 11.15 Air flow through a windcatcher (windtower/Badgir) (Tavassoli 2002)

11.3.1 Shade and Rhythms in Urban Morphology

Creating shade is one of the main elements of adaptation to hot-arid climate in Persian architecture and cities. Shaded areas not only protect the people from the hot rays of sun in winter, but also by creating cooler areas, in contrast to the unshaded areas, create winds which cool the environment as well. The climatic comfort in the neighborhood is produced by arches connecting two sides of the alley wall called: “Sabats”.

An example is the rhythm of Sabats in Yazd city which not only create light and dark spaces, which are aesthetically pleasant to the eye, but also rhythms of shade between the spaces act as cooling agents. Part of shade creation is achieved through the proportions of public spaces which is usually 1:1 to 1:3.5 (Karami and Mohammadi 2022) (Fig. 11.15).

11.3.2 Shade

Creating shade by building arches in the walls as functional ornaments or creating seating places for spectators in religious and national ceremonies and festivals such as Ashoora. As an example Amir Chakhmagh Square in Yazd has rows of arches in the enclosing wall of the square, creating spectacular places especially for women to be separated from the ceremony which is held by men (Fig. 11.16).

Fig. 11.16 Creating shade on façade by arches in AmirChakhmagh Square, Yazd. (https://wikibin.ir/article/wp)

Fig. 11.17 Yazd Sabats in a typical alley. (https://api.saanaa.com)

11.3.3 Alley and Walkways

To protect people from the heat, streets and sidewalks are constructed mainly in the east–to-west direction. On hot summer days, the narrow alleys with high walls on both sides are completely in the shadows (Tavassoli 2002). To prevent the hot air from flowing down the streets and sidewalks and entering the buildings, the sidewalks are mostly curved and narrow with high walls. Roofed sidewalks (Sabat) help keep the sidewalks in the shadows, and narrow sidewalks contribute to the density of nearby buildings (Fig. 11.17).

11.3.4 Facades and Fenestration

The materials used for plaster are raw clay, and a mixture of clay and hay (Kahgel) to coat the walls (Moradi 2005; Keshtkaran 2011). Their light colors and textures absorb less heat and reflect light. The numbers of windows that open to the sidewalks are minimized to avoid having the unfavorable climate penetrate the indoor area. To prevent dust and excessive sunlight from getting into the buildings, windows and openings are usually placed in the ceiling or high up the walls. Most of the windows open to the protected central courtyard area, which generally has less harsh conditions and a more favorable environment than exists on the outside of the buildings (Meamarian 1999).

The windows in hot and dry regions are generally small and are in the upper parts of walls just near the ceiling. Although external walls do not have many windows most of them on the yard facing internal walls. Passive ventilation is done by these windows. Windcatchers also help to the internal ventilation. (A'zami et al. 2005).

11.3.5 Use of Water: Kariz/Qanat in Urban Morphology

The Kariz or Qanat is an irrigation system which leads underground water from mountain foots to the middle of deserts. The waterways are either underground or are designed to be accessed on the ground. The forms of access can be through (underground or on-ground) streams or ponds or Payab (direct access point to the Qanat).

One of the examples of how the Qanat network has affected the urban morphology is studied by Soltani Mohammadi (2018) in the Mohammadieh neighborhood in Naeen city in central Iran. The Qanat network has determined the morphology of the street network and the location of the main public functions as well as the direction and location of houses along the streets. There are two types of Qanats in Mohammadieh, one is agricultural Qanat (called Hanafsh) which enter the farmlands as soon as comes up to the surface. The other is the urban-farming Qanat which enters the city. After irrigating the private and public buildings, it enters the orchards surrounding the residential areas. The Qanat route affects the structure of city which is also affected by the topography, natural setting, the organic and compact morphology of the city, the access ways and street network. The location of public spaces and residential buildings also follow the Qanat network. To use water in architectural scale the Qanat network is connected to public baths, the neighborhood center, mosques, alleys and a mortuary. In each of the buildings there is a place called "Keh" which is the access point of the building to the Qanat access way.

11.4 Conclusion: Design Guidelines Derived from Persian Architecture, Design and Planning in Hot Arid Climates

As the world is suffering from increasing effects caused by disasters such as extreme heat, droughts and floods, there is a rising need for new approaches, and methodologies to (re)design the urban and rural environments to keep the built environment livable. We need to study the methodologies and approaches used in Iranian architecture and urban design traditions to relearn and reuse their adaptation methods to achieve maximum liveability and comfort in extreme heat and scare water resources though natural and energy sufficient ways. This chapter has studied the traditional knowledge of building and design of Iranian cities and buildings in hot dry conditions to be inspirated by practices of millennia and centuries ago and implement these in our contemporary ways of building and planning.

This chapter has introduced the principles of Persian architecture and planning from microscale of building materials, to building design and architecture, the neighborhood and urban morphology and planning in larger scales.

11.4.1 Creating Shade

One of the most important lessons learned from Persian traditional architecture and urban design and planning is vast use of shade which are functional in several ways.

1. Creating spaces which are protected from direct sunlight and are several degrees cooler than the exposed areas.
2. Creating a rhythm of shade and light: warm and cooled surfaces which cause are movement and wind which in turn helps with naturally cooling the areas.
3. Using underground spaces as exposed: Godalbaghche/pit yard or Sardab/basement as spaces which have maximum shade and coolest.

11.4.2 Sustainable Use of Water

Water as the main element for livelihood and survival is also used for passive cooling in the buildings. The Qanats have created sources of water where nature does not provide any water sources, their presence on the surface is where the public and private buildings are located. Acceding the water from the Qanat is where the core of the building is shaped, underneath or over the surface, inside the building or in the courtyards. In addition, the water is used as an additional tool for cooling the wind coming from the Badgir.

11.4.3 Vegetation

Creating microclimates by planting local flora can create paradises inside the courtyards. The shade and pleasant presence of the plants create thermal and climatic comfort by increasing the environmental humidity.

11.4.4 Passive Use of Wind, Sunlight, and Orientation

The traditional Persian architecture has made the best use of natural elements including the wind and the sun and to do so, the orientation of the buildings according to the location of the sun and angle of the sunlight in winter and summer. Moreover, the direction and intensity of the dominant wind and the temperature of each wind direction is carefully studied and used in designing the buildings and choosing their location in the lot. In addition, the effects of sun and wind are considered in the morphology of the neighborhood and city. The way buildings are set in the larger scale is chosen to reduce the harsh climatic conditions such as reducing the contact with the direct sunlight in summer and harsh winds. This is done by densifying the context, attaching the buildings and carving out inner courtyards which give the residents maximum protection from the hard climatic conditions.

11.4.5 Using Different Building Levels for Maximum Climatic Comfort

Using the climatic potentials of different levels of underground, ground level and above the ground has helped to create maximum comfort for people in a passive way. By using the underground, this creates more shade and gives access to cooler layers of the earth and Qanat channels. Many of the subsoil spaces such as Sardab have been used for many centuries and were an inseparable part of the traditional buildings even in milder climates. Using the rooftop to catch the passing breeze by Badgir (one sided, two sided or four sided) plays a great role in thermal comfort of the buildings. Badgirs are one of the main elements of Persian architecture which can be found at the far end of Persian Empire and Silk Road via trade and wars.

Revival of traditions and lesson learned from the wisdom of the past where old civilizations found ways to adopt to climatic and geographical conditions of their land, is not only wise, but helps us localize the global solutions according to the millennia of culture and knowledge. The Persian architects and designers have built their buildings and cities, in ways that not only maximized their thermal and environmental comfort but also created beauty, function and cultural values. The design guidelines derived from Persian architectural design and planning in hot arid climates include the use of elements which can be replicated and derived in new ways in contemporary architecture, design, and planning.

References

A'zami A, Yasrebi SH, Salehipoor A (2005) Climatic responsive architecture in hot and dry regions of Iran. In: International conference "passive and low energy cooling for the built environment", Santorin, Greece

Asghari-Moghadam M, Rajabi A (2006) Natural geography of the city: cities of dry areas. Sara Publications, Tehran. (in Persian)

Baghai P, Ansari M, Bamanian MR, Fayaz R (2014) The range of thermal comfort in the traditional residential open space of Yazd city. Hoyt Shahr Mag 23. Fall

Bahrpeima S, Sargazi M (2022) Climatic Design, adaptation with environment: case study, hot arid climate of Iran. Res Urban Plan 13(49):282–292. https://doi.org/10.30495/JUPM.2022.5530

Bemani MR (2002) Introduction to the role and application of peymoon, Art and Architecture, Issue 1

Darani M, AliAskarian H, Salehi SM (2013) Fuel consumption in building (case study west of Isphahan, Second national Climate conference, held by SABA

Ebadi H, BigdeliRad V, Bin Lamit H, Yegane S, Kermaji R (2014) A survey on sustainability of central courtyards of Iran traditional architecture. World Appl Sci J. https://doi.org/10.5829/idosi.wasj.2014.30.icmrp.29

Encyclopaedia Iranica (1989) iranicaonline.org

Eskandari P (2011) Analysis of traditional Iranian houses of Kashan, Iran in Terms of Space Organization and Access Design. North Cyprus, Eastern Mediterranean University, (5) (PDF) Iranian courtyard housing: the role of social and cultural patterns to reach the spatial formation in the light of an accentuated privacy

Farrokhyar H, Karimian H (Translator), Modarres Zadeh A (Editor) (2014) One hundred houses, one hundred plans (features of old houses' architecture in historical contexts–hot and dry climate). Islamic Azad University (Kashan)

Foruzanmehr A (2014) Thermal comfort and practicality: separate winter and summer rooms in Iranian traditional houses. Archit Sci Rev 59:1–11. https://doi.org/10.1080/00038628.2014.939132

Gharehgolchian A (2010) Concept of solar sustainable heat and cool. The first national conference of contemporary architecture & urbanization. Shiraz, Iran

Ghobadian V (1998) Climatic analysis of the Iranian traditional buildings. Tehran University Publications, Tehran

Ghobadian V, Mahdavi M (1993) Climatic design. Tehran University Press, Tehran. (in Persian)

Golkar K (2000) Sustainable urban design in the sidelines of desert. J Honar-ha-ye-ziba Memari-va-Shahrsazi 8:43–52. (in Persian)

Hosseini SR, Nik Ethegad A, Uson E, Armesto A (2015) Iranian courtyard housing: the role of social and cultural patterns to reach the spatial formationin the light of an accentuated privacy [en línea] Fecha de consulta: dd-mm-aa. En: ACE: Architecture, City and Environment = Arquitectura, Ciudad y Entorno 10(29):11–30. https://doi.org/10.5821/ace.11.29.2653. ISSN: 1886-4805

Jafarian S, Monsefi Parapari D (2021) Role of the central courtyard in hot arid climate: A case study of Tabatabai and Borujerdi houses in Kashan, Iran

Karami H, Mohammadi M (2022) An analysis of the pattern language of historical squares in the Yazd region and the development of urban square design guidelines. J Archit Hot Dry Climate 9(14):181–202. https://doi.org/10.29252/ahdc.2022.16431.1538

Keshtkaran P (2011) Harmonization between climate and architecture in vernacular heritage: a case study in Yazd, Iran. Procedia Eng 21:428–438

Kiani MY (2004) History of Iranian architectural art during the Islamic era: the organization for researching and composing university textbooks in the humanities (Samt). Dehkhoda, "Dictionary", Tehran

Mahmoodi M (2009) The Badgir. Yazda, Tehran, pp 25–40

Meamarian GH (1999) Residential architecture of Iran. The University of Science & Industry, Tehran, pp 239–347

Monsefi Parapari, D. (2015). Adaptation to climate change and thermal comfort. TU Dortmund, Germany

Moradi S (2005) Environmental control system. Ashiyan, Tehran. pp 34–41, 55–57, 128–134

Nayebi F (2002) Life in the courtyard. Nezhat Publications, Tehran

Piryaei M, Shemirani SMM, Sabernjad J (2022) Morphological investigation of the central courtyard of hot and dry local buildings; Case study: Yazd City. Armanshahr Archit Urban J 15(40):31–41. https://doi.org/10.22034/AAUD.2021.265165.2388. http://www.armanshahr-journal.com/article_163871.html

Pirnia MK (1993) Islamic architecture of Iran, 2nd edn. TehranDaneshgah-e Elm va Sanat Publications

Pirnia MK (1999) Typology of Persian architecture styles. Soroush Dansh publications

Roaf S (1982) Wind catchers: living with the desert. (Beazley E, ed) Air and Philips, England

Salehpour H, Ali R, Javad R (2013) The effective effect of water on architectural sustainability of hot and dry climate homes, national conference on architecture, urban planning and sustainable development with a focus on vernacular architecture to sustainable city, riki_ali_20@yahoo.com

Saljoughinejad S, Sharifabad SR (2015) Classification of climatic strategies, used in Iranian vernacular residences based on spatial constituent elements. Build Environ 92:475–493

Shamlou N (2007) Investigation of the role of water and soil in the sustainability of desert housing. In: Proceedings of the second scientific conference on sustainable architecture, Tehran

Soflaei F (2012) Stability of climatic elements in traditional architecture of Iran (hot and dry climate), the third conference on optimization of fuel consumption in buildings, Tehran. https://civilica.com/doc/2457

soflaei F, Shokouhian M (2008) The environmental impact of courtyards in Iran's sustainable architecture (hot and dry regions, climate), (case study: courtyard houses in Isfahan and Kerman). J Urban Manag Stud. https://ums.srbiau.ac.ir/article_22995.html?lang=en

Soltani Mohammadi M (2018) The impact of Qanat in physical form case study: Mohammadiyeh Village, Nain, Iran. J Architect Hot Dry Clim 5(6):97–115

Soltanzadeh H, Sadatzad G (2013) Role and use of water and courtyards in sustainability of traditional Iranian architecture; architecture and sustainable development conference, Bookan (in Persian)

Tavassoli M (2002) City building and architecture in hot and dry climates. Payam Publications Tehran. (in Persian)

Tayari N, Nikpour M (2022) Investigation on daylight quality of central courtyard's adjacent rooms in traditional houses in hot dry region of Iran: a case study Yazdanpanah house. Iran J Energy Environ 13(4):320–332. https://doi.org/10.5829/IJEE.2022.13.04.01

Vakilinezhad R, Mofidi SM, Seraj FM (2013) Shanashil: a sustainable element to balance light, view, and thermal comfort. Int J Environ Sustain 8(4):101–110

ZadehEsmaeel N, TorabiLangari M (2013) Role of water in sustainability of Hot Arid climate architecture, case study: Kerman old urban houses; architecture and sustainable development conference, Bookan (in Persian)

Zangeneh N, Sepidkar H, Eghtesadi N (2014) Comparative analysis of the impact of climate on the geometric shape of the Shabestans of mosques in shiraz and Tabriz. In: National conference on architecture, civil engineering, and modern urban development, Tabriz. Retrieved from https://civilica.com/doc/315133

Zareei, ME, Mirdehghan SF (2016) The influence of the central yard in moderating environmental condition of the hot and dry climate of Yazd region, J Stud Iranian Islam City, 6(23), 5-18

Zarei H, Razani M, Ghezelbash E (2017) Reconstructing the designs pattern of shiraz historical houses approaching climate in the Qajar period. Pazhohesh-ha-ye Bastanshenasi Iran 7(13):225–242

Zeynalian N, Okhovat H (2017) Morphology of courtyard in Qajari houses in hot-dry and hot-humid climates, with a focus on the species of "courtyard in the middle" (case study: Yazd and Dezful houses). J Stud Iranian-Islamic city 8(30):15–29

Image References

https://www.isna.ir/photo/96070904524/%D9%85%D8%B1%D8%A7%D8%B3%D9%85-%D9%86%D8%AE%D9%84-%D8%A8%D8%B1%D8%AF%D8%A7%D8%B1%DB%8C-%D8%B8%D9%87%D8%B1-%D8%B9%D8%A7%D8%B4%D9%88%D8%B1%D8%A7-%D8%AF%D8%B1-%D9%85%DB%8C%D8%AF%D8%A7%D9%86-%D8%A7%D9%85%DB%8C%D8%B1-%DA%86%D8%AE%D9%85%D8%A7%D9%82-%DB%8C%D8%B2%D8%AF
https://fa.wikipedia.org/wiki/%D9%85%D8%AC%D9%85%D9%88%D8%B9%D9%87_%D9%85%DB%8C%D8%AF%D8%A7%D9%86_%D8%A7%D9%85%DB%8C%D8%B1%DA%86%D8%AE%D9%85%D8%A7%D9%82
https://api.saanaa.com
https://utravs.com/umag/Iran-Attractions/amir-chakhmaq-complex
https://en.wikipedia.org/wiki/Abb%C4%81si_House
https://api.saanaa.com/cdn/download/2/1400/07/20/%D8%B3%D8%A7%D8%A8%D8%A71-637696316675174915.jpg
https://wikibin.ir/article/wp/%D9%85%DB%8C%D8%AF%D8%A7%D9%86_%D8%A7%D9%85%DB%8C%D8%B1_%DA%86%D8%AE%D9%85%D8%A7%D9%82
Image 158104 - UNESCO World Heritage Centre
climatology.ir
https://www.japantimes.co.jp/news/2023/07/28/world/iran-wind-catcher-cooling-architecture

Chapter 12
Evapotranspiration Impacts on Summer Surface Urban Heat Island Distributions and Trends in Santa Clara, California: Moderating Effects of the Human Environment

Ana Rivera, Nathan Moore, Jun Hyun Kim, Sue C. Grady, and Robert D. Bornstein

Abstract Increased urbanization impacts surface albedos and alters surface energy balances, reducing urban evaporative cooling and thus forming surface urban heat islands. Neighborhoods with higher vegetative densities, therefore, tend to have lower temperatures than those with sparse vegetation. In the United States, racial and ethnic segregated neighborhoods have limited green infrastructure, exposing populations to higher ambient temperatures. This study determined seasonal daytime and nighttime land surface temperature (LST) trends from MODIS data and distributions from ECOSTRESS data in Santa Clara County, California. It related these results to vegetation (NDVI) and evapotranspiration (ET) values, as well as to median household income, percentage of Hispanic/Latino populations, and the built environment per the Local Climate Zone (LCZ) classification framework. The combination of these assessments comprised a comprehensive human-environment approach for heat exposure evaluation. Results show upward trends in daytime summer and winter mean LST and weak nighttime trends. Winter NDVI and LST values exhibit positive correlations. Negative correlations are found for summer NDVI and

A. Rivera (✉) · N. Moore · S. C. Grady
Department of Geography, Environment and Spatial Sciences, Michigan State University, East Lansing, MI, USA
e-mail: rivera59@msu.edu

J. H. Kim
School of Planning, Design, and Construction, Michigan State University, East Lansing, MI, USA

R. D. Bornstein
Department of Meteorology and Climate Science, San Jose State University, San Jose, CA, USA

NOAA Cooperative Science Center for Earth System Science and Remote Sensing Technologies (CESSRST), The City College of New York, New York, NY, USA

R. Roggema (ed.), *Dry Urbanism*, Contemporary Urban Design Thinking,
https://doi.org/10.1007/978-3-031-77347-1_12

LST values, which are stronger during daytime hours, indicating that the cooling effects of vegetation occur primarily during the daytime. Maximum LSTs occur in low-income neighborhoods characterized by high densities and scarce vegetation with a high percentage of Hispanic/Latino populations.

Keywords ECOSTRESS · Land surface temperature · Evapotranspiration · Thermal inertia · Race and ethnicity

12.1 Introduction

Population growth has driven an increase in anthropogenic greenhouse gas emissions into the atmosphere through the combustion of fossil fuels (IPCC 2014). These also affect local climate conditions in cities, where anthropogenic releases of heat, moisture, and pollutants and changes in surface properties are some of the leading causes of modified urban climates (Oke 1987). Heat exchange between the surface and air depends on the thermal properties of the surface and atmospheric conditions (Oke 1982). Urbanization impacts surface albedos and evapotranspiration (ET) rates, altering surface energy balances and thus creating distinctive local urban climates, including the formation of urban heat islands (UHIs; Oke et al. 2017). UHIs negatively impact energy use, air quality, and human thermal comfort and health (He 2018). Children, the elderly, low-income groups, and people with preexisting respiratory and cerebrovascular conditions are most vulnerable to these effects (McMichael 2000).

UHIs are stronger at the surface (Duckworth and Sandberg 1954; Bornstein 1968) under calm wind and clear sky conditions, and they decrease in magnitude outward from the urban core (Raj et al. 2020). Their intensity, defined as the difference between urban and rural temperatures (ΔT_{u-r}),exhibits temporal variations due to several factors (Peng et al. 2012), including differences between urban and rural vegetation, resulting in differing ET cooling rates (Jauregui 1997). The thermal inertia, the degree of slowness with which the temperature of a body approaches that of its surroundings (Ng et al. 2011), is higher for wet rural areas than for urban environments, and thus wet-rural areas (vs. urban areas) warm more slowly during daytime hours and cool more slowly during evening hours (Swaid 1991); the reverse is true for dry rural soil areas. Daytime UHIs are thus, in theory, stronger, and nighttime values are weaker for cities surrounded by wet rural environments; the reverse is thus true for cities surrounded by dry rural soils (Imamura 1991; Bornstein et al. 2012; Dou et al. 2015).

UHIs can be estimated via 2-m air temperatures measured by fixed or mobile sensors, producing canopy UHI values. Measurements of canopy UHIs are, however, limited by network densities, so some studies use satellite-derived land surface temperature (LSTs) to estimate surface UHIs magnitudes and distributions (Voogt 2007; Stewart and Mills 2021). Previous results from the Moderate Resolution Imaging Spectroradiometer (MODIS) sensor report negative annual-mean daytime

MODIS surface UHI intensity trends over 2003–2016 were found in arid Lima, Peru (Wu et al. 2019), as rural warming rates were larger than in urban areas. Positive trends in the extent and intensity of surface UHI over 17 years were found in tropical-wet Chennai, India (Raj et al. 2020). A similar positive annual trend was also found in Pune, India, which means its wet season trend is stronger than its dry season value. Negative trends in nighttime values were observed in tropical-wet Lucknow, India (Siddiqui et al. 2021). The same study also reported negative daytime trends, contrary to the generalization of Swaid (1991). Another contradictory study is from tropical-wet Cali, Colombia, which showed increased annual nighttime surface UHI intensity from 2003 to 2016 (Wu et al. 2019).

An analysis by Li et al. (2020) of large cities in various climates showed 2003–2013 MODIS-derived daytime surface UHIs amplified by increased wet seasonal precipitation while corresponding nighttime values were reduced or became negative, as wet season rural moisture is greater than urban values. Other examples include MODIS-derived weak or negative daytime surface UHIs in dry conditions, e.g., during the 2001–2018 dry season periods in Pune, India (Siddiqui et al. 2021) and from irrigated urban parks in arid Phoenix, Arizona, USA, during 2003–2005 (Imhoff et al. 2010). In Mediterranean climates (e.g., Csa in the Köppen classification system), daytime surface UHI intensity peaks in late spring and early summer and decreases as precipitation declines (Sismanidis et al. 2022). As rural vegetation dries due to reduced precipitation, ET decreases, and so does the surface UHI intensity. The UHI intensity is, therefore, directly linked to local precipitation (Manoli et al. 2020).

The reduction of evaporative cooling is an important factor contributing to UHIs, as urban moisture content is lower than in non-urban areas, allowing urban areas to warm faster than their vegetated surroundings (Taha 2017). Droughts, however, can intensify UHIs as insufficient soil moisture in urban areas leads to less latent heat by evapotranspiration (Dong et al. 2018). During the 2012–2016 drought in Los Angeles County, USA, ET values, measured with the combination of Landsat and MODIS, declined from 29.3% in 2011 to 24.6% in 2016, and annual mean daytime LST increased from 34.4 to 37.1 °C as the cooling effect from grasses was reduced due to physiological responses, irrigation changes, and losses of green vegetation cover (Allen et al. 2021). The Normalized Difference Vegetation Index (NDVI), a measure of vegetation cover, has been used to monitor drought conditions and to extract information on soil moisture (Thanabalan et al. 2023). NDVI values range from −1 to 1, with negative values indicating water, positive values near zero indicating bare soil, and higher positive values ranging from sparse vegetation (0.1–0.5) to dense green vegetation (0.6 and above). In Los Angeles, CA, MODIS-derived NDVI and LST exhibit strong negative correlations, showing that the availability of vegetation may enhance local cooling (Tayyebi and Darrel Jenerette 2016).

As droughts continue to impact the State of California, limits to outdoor watering will reduce ET rates in urban areas, decreasing their cooling capacity and increasing UHI intensities. Within urban areas, however, outdoor watering varies significantly. High-income neighborhoods, for example, are strongly correlated with highly irrigated areas, as high-income households are willing to pay for irrigation despite irrigation limits (Miller et al. 2022). In Los Angeles, higher income was related to greener landscapes, whereas less green landscapes were related to higher percentages of Hispanic or Latino residents (Mini et al. 2014). There is a need to further understand the role of ET and its relation to surface UHI intensity, particularly in areas where spatial variations in ET may partly depend on water usage differences among residents of varying income levels.

UHIs have shown significant intra-city variability due to the heterogeneity of urban areas (e.g., differences in the intensity of human activities, amount of green space, and amount and properties of built-up materials), suggesting higher temperatures located in neighborhoods inhabited by low-income populations (Chakraborty et al. 2020). Intra-urban surface UHI variability also exhibited seasonal and spatial variations resulting in higher intensity during summer days and within high-density zones (Wang et al. 2023). The intensity of surface UHIs, additionally, has high temporal and spatial variability (Rivera et al. 2017), rendering some sensors insufficient to understand these variabilities, e.g., MODIS has a high-temporal (1–2 days) but a low-spatial resolution (1 km), while Landsat 8 has a high-spatial (100 m) but a low-temporal resolution (16 days). ECOsystem Spaceborne Thermal Radiometer Experiment on Space Station (ECOSTRESS), aboard the International Space Station, provides daily coverage (superior to Landsat) at a 38 by 69-meter resolution (superior to MODIS), with higher latitude areas measured multiple times per day.

ECOSTRESS data can provide daily LST and ET values; the latter is a valuable indicator of consumptive water use, as it results from the direct evaporation of water and plant transpiration (Cawse-Nicholson et al. 2021), allowing for the understanding of ET and LST response at the intraurban level. An analysis with the use of ECOSTRESS by Coleman et al. (2020) in Los Angeles shows that LSTs for irrigated and non-irrigated areas were similar in the morning (8–11 a.m.) when atmospheric conditions favor stomatal conductance in both regions. LST values were, however, different in the afternoon (1–3 p.m.), as high temperatures can cause heat stress in non-irrigated areas as soil moisture is insufficient to meet atmospheric demands (Coleman et al. 2020). ECOSTRESS data, therefore, provide the capacity to analyze surface UHI variations at high temporal and spatial resolution (Chang et al. 2022). The analysis of long-term LST trends, however, requires large data sets, which limits the number of thermal images that can be analyzed (Ermida et al. 2020). Additional problems arise from inconsistent definitions for land-use and land cover classes, a challenge for comparison studies.

The Local Climate Zones (LCZs) classification system (Stewart and Oke 2012) was developed to standardize urban morphology classes. It recognizes 10 urban and seven non-urban LCZ areas with uniform surface cover, structure, construction materials, and human activity. LSTs vary broadly by LCZ class, with maximum LST usually recorded for compact and heavy industry areas (Cai et al. 2018). In conjunction with related seasonal variations in urban vegetative cover, soil moisture variation can alter the typical LST-LCZ relationships (Ren et al. 2019). High-density urban areas in compact mid- and low-rise classes have high-heat stress zones (Verdonck et al. 2018) due to their high population densities. Even though industrial and paved classes exhibit high LSTs, they produce only minor stress impacts due to their low densities.

Whereas most previous studies have primarily focused on the LST spatial variability among LCZs, few have explored the use of the LCZ framework to analyze thermal inequality. The current study, therefore, uses a combination of remotely sensed data to determine temporal and spatial LST distributions and trends in Santa Clara County, California, and its relationship to the human environment. A 20-year summer averaged MODIS data was used to determine annual winter and summer LST and NDVI temporal trends. ECOSTRESS-derived summer-mean LST and ET spatial variability, and its relation to LCZs was then studied. The association between race and LCZs, as well as income and LCZs, were then explored via multinominal logistic regressions. The results identified Hispanic/Latino and low-income populations at potentially high heat stress risk from a combination of environmental and social factors, which allowed for the investigation of nature-society relationships: the role of humans in changing the environment and how this changed environment influences humans (Moseley et al. 2013).

12.2 Data and Methods

The first step in this framework evaluated the physical environment via the calculation of annual and seasonal daytime and nighttime LSTs and their corresponding NDVI values from MODIS for 2000–2020 to determine temporal trends. A four-year summer LST and ET ECOSTRESS for 2018–2022 was then calculated to determine spatial trends. The second step evaluated the social environment per the LCZ framework and US Census data to describe the built environment and assess the distribution of Hispanic/Latino and low-income populations across census tracts in Santa Clara County. The third step analyzed summer LST and ET in relation to LCZ and LCZ in relation to socioeconomic factors. The combination of these assessments resulted in a comprehensive human-environment approach to evaluate LST intensity in areas of varying levels of median household income in Santa Clara County.

12.2.1 Study Area

The study site is Santa Clara County (Fig. 12.1), located in the San Francisco Bay Area (37°18′23”N, 121°54′46”W). Santa Clara County covers an area of 3380 km^2, making it the 37th largest county in California. The study area is surrounded by water from the San Francisco Bay to the North, the Santa Cruz Mountains to the southwest, and the Diablo Range to the northeast. According to the National Land Cover Dataset, 25% of Santa Clara County was classified as urban developed. The rest of the study area was comprised of forest (31%), shrubland (24%), and

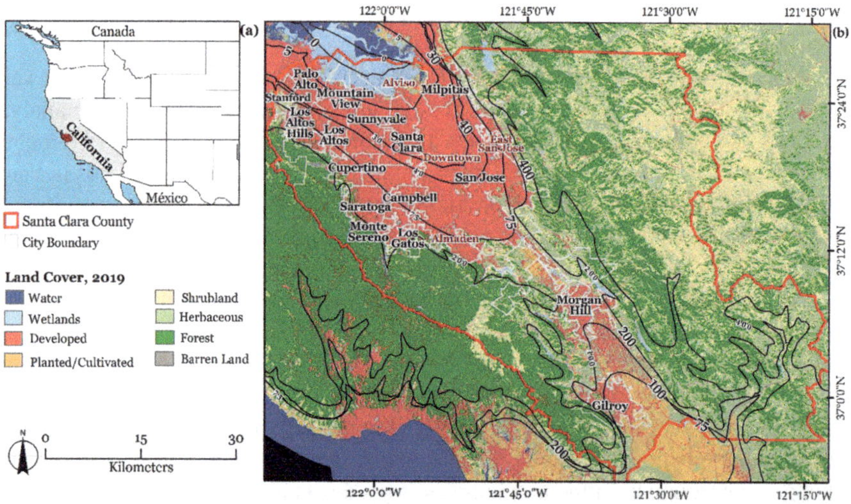

Fig. 12.1 Location of Santa Clara County: (**a**) in Northern California (red area) and (**b**) its land cover (USGS 2019), 15 cities, and historic neighborhoods of Alviso, East San Jose, and Almaden, as well as topographic contours (meters above mean sea level)

herbaceous vegetation (14%). Only 2% of the non-developed area was classified as planted or cultivated. The remaining area was comprised of water (2%), wetlands (<1%), or barren land (<1%). The US Census data (2020) reports that the study area had a population of 1.9 million, an increase of 0.15 million from 2010.

12.2.2 Physical Environment

MODIS. Annual summer and winter LST and NDVI trends were determined from MODIS. A total of 14,303 satellite images from 5 March 2000 to 31 December 2020 were accessed for Santa Clara County to derive 21 daytime (8 a.m.–12 p.m) and 21 nighttime (8 p.m –12 a.m.) monthly mean LST trends, as well as the corresponding annual mean NDVI values. For seasonal analysis, summer is considered from 1 June to 31 August, whereas winter is considered from 1 December to 28 February. Images were retrieved via Earth Engine (code.earthengine.google.com), which provides free access to satellite datasets (Gorelick et al. 2017), allowing for long-term LST and NDVI trend computations (Budhiraja et al. 2020; Chakraborty and Lee 2019; Wang et al. 2020). Cloud-covered pixels (>0% cloudiness) in MODIS images were first masked using the Quality Assessment (QA) product that provides information on pixel quality and allows users to apply pixel-based filters. The National Land Cover Dataset (USGS 2019) was then used to retrieve pixels corresponding with values classified as urban developed (classes 21–24) for each LST and NDVI satellite product. This process filtered non-developed (rural) areas from developed (urban) locations to better estimate the urban variability of LST values (Rivera et al. 2022).

ECOSTRESS. Approximately 300 ECOSTRESS scenes were retrieved via the NASA Earth Data Cloud web application (appeears.earthdatacloud.nasa.gov) for the summer months (June to August) of 2018 to 2022. Determination of LST (Level-2) and ET (Level-3) distributions first involved a per-pixel filter applied to ECOSTRESS images, using its cloud product to mask cloudy and cloud shadow pixels. As with the MODIS products, pixels corresponding with values classified as urban developed (classes 21–24) for each LST and ET satellite product were retrieved. These values were combined to produce summer mean LST (°C) and ET (Watts/m^2) distributions for the study area.

12.2.3 Socioeconomic Environment

Local Climate Zones. The built environment of Santa Clara County was described in the current study via the LCZ framework of Stewart and Oke (2012). The LCZ classification system, developed to standardize urban morphology classes, recognizes 10 urban and seven non-urban LCZ areas with uniform surface cover, structure, construction materials, and human activity. This methodology is based on the World Urban Database and Access Portals Tools (WUDAPT) of Ching et al. (2018) and integrates satellite imagery with open-source software (Bechtel et al. 2015).

User knowledge of local morphological conditions helped create the minimum required 20 WUDAPT training samples for each LCZ class via Google Earth software, using 20 June 2020 imagery as the base layer. Less than 15 training samples were digitized for LCZ 2 (compact mid-rise) as it is rare in the study area, and no LCZ 1 (compact high-rise) and LCZ 4 (open high-rise) areas exist in the study area. The samples were then used as input to run an LCZ classification algorithm in the LCZ Generator (lcz-generator.rub.de), a web application to create LCZ maps with its corresponding automated accuracy assessment (Demuzere et al. 2021). As with previous datasets, pixels corresponding with values classified as urban developed (classes 21–24) were retrieved.

Neighborhood demographics. Racial/ethnicity and economic data were acquired at the census tract level for Santa Clara County from the US Census Bureau, American Community Survey Five-Year Estimates (2017–2021). Since the Hispanic/Latino population comprises a third of the total population in the County, this ethnic group was selected to explain the moderating effects of race on environmental outcomes. The Hispanic/Latino (DP05_0071E) variable was divided by the total population variable (DP05_0001E) to estimate the percentage of the Hispanic/ Latino (any race) population per census tract. The socioeconomic status of residents was measured with the use of the median household income variable (DP03_0062E). Census Tract 5116.08 was omitted as it represented Stanford University and had no income data.

12.2.4 Human Environment

The Zonal Statistics tool in ArcGIS Pro© (version 3.0.3) was used to calculate the mean LST and ET values per census tract. The potential association between the LST, ET, and percentage of Hispanic/Latinos, as well as median household income, was explored with Spearman correlation statistics, as values do not follow a parametric distribution. The most frequently occurring LCZ class per census tract was first found with the use of the Zonal Statistic Tool, majority option, and assigned to each spatial unit. The relationship between the built environment per the LCZ classification system and race/ethnicity, as well as median household income at the census tract level was examined with the use of descriptive statistics displayed in boxplot diagrams.

12.3 Results

The human-environment conceptual framework allows for the study of the temporal and spatial variation of LSTs related to NDVI and ET values, as well as Hispanic/Latinos in Santa Clara County. MODIS-derived LST results show stronger daytime summer (0.10°C per year) and winter (0.11°C per year) means than their corresponding nighttime values (0.03 and 0.04°C per year, respectively). MODIS-derived NDVI results show no strong temporal trends for summer and winter. Positive correlations were, however, found for winter NDVI and LST values, whereas negative correlations were found for corresponding summer values.

ECOSTRESS-derived 2018–2022 summer LST values for the entire domain have a mean of 27.11°C with a corresponding ET mean of 129.73 Watts/m^2. The spatial distribution analysis shows an increasing LST with an increasing percentage of Hispanic/Latino populations (0.06°C per percent increase in Hispanic/Latinos). Hispanic/Latinos in Santa Clara County are, therefore, most likely to experience high LST values due to low soil moisture resulting from high-density buildings (compact low-rise) and the prevalence of impervious surfaces in large low-rise areas.

12.3.1 Physical Environment

MODIS. Annual-mean summer LST from 2000 to 2020 (Fig. 12.2) shows a slight upward trend for daytime values (m = 0.10°C per year, R^2 = 0.52), with a minimum of 37.37°C in 2001 to a peak of 40.77°C in 2020, along with large yearly annual variabilities (±1σ) ranging from 2.93 to 3.33°C. The corresponding nighttime values (Fig. 12.2b) show a weak upward trend (m = 0.03°C per year, R^2 = 0.14) with a minimum of 14.60°C in 2011 to a maximum of 16.33°C in 2017, and smaller annual variabilities (1.11 to 1.44°C).

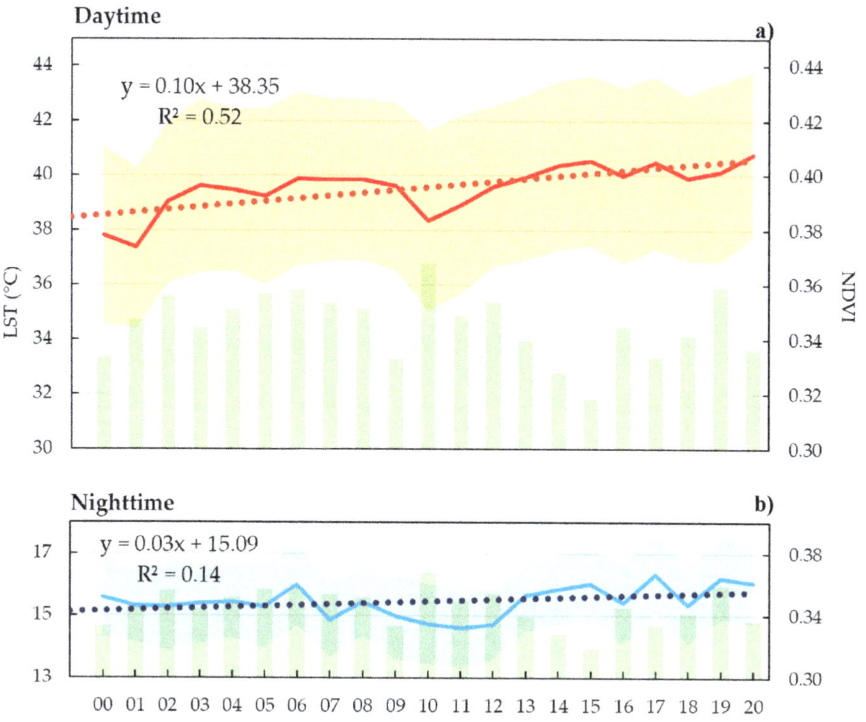

Fig. 12.2 MODIS (2000–2020) annual-mean and ± 1σ (color band) variations of summer (**a**) daytime and (**b**) nighttime LSTs (°C), as well as annual mean NDVI values

Annual summer LST variations are slightly larger for daytime (0.39°C) than for nighttime (0.33°C) hours due to daytime turbulence levels. Annual-mean summer NDVI values range from 0.32 in 2015 to 0.37 in 2010 and exhibit no trend (m = −0.0005 per year, R^2 = 0.08). The Spearman correlation between summer annual-mean LST and NDVI shows a negative but moderate relationship for daytime (rho = −0.34, p-value = 0.13) and nighttime (rho = −0.43, p-value = 0.05) values.

MODIS annual-mean winter LST values (Fig. 12.3) show a slight upward daytime trend (m = 0.11°C per year, R^2 = 0.28), similar in magnitude to the daytime-summer LST trend. The values range from 15.03°C in 2001 to 19.67°C in 2014 and exhibit small yearly annual variabilities (±1σ), ranging from 1.36 to 1.98°C. Corresponding nighttime annual-mean LSTs (Fig. 12.3b) show a weak upward trend (m = 0.04°C per year, R^2 = 0.07) with values ranging from 3.18°C in 2006 to 7.16°C in 2014, and very small annual variabilities (0.58 to 1.06°C).

Annual LST variations are larger for daytime (0.62°C) than nighttime (0.48°C) hours, and both values are larger than their corresponding summer annual measures. Annual-mean NDVI ranges from 0.18 in 2009 to 0.27 in 2017, showing a larger range than their corresponding summer values. Winter NDVI exhibits a non-existent trend (m = 0.002 per year, R^2 = 0.25). The Spearman correlation between winter annual-mean LST and NDVI shows a strong and significant positive relationship for

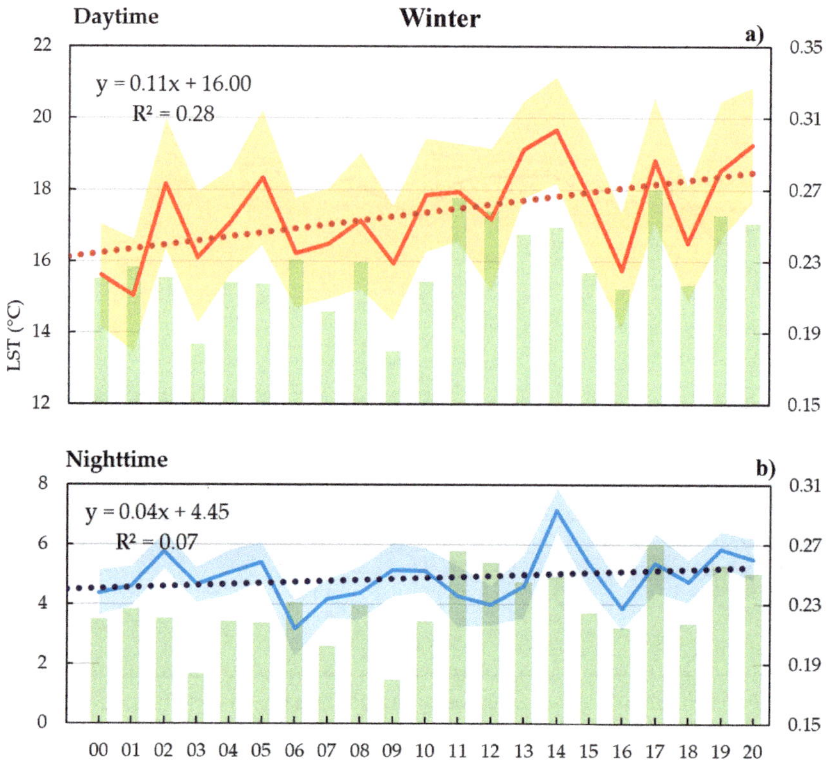

Fig. 12.3 MODIS (2000–2020) annual-mean and ± 1σ (color band) variations of winter (**a**) daytime and (**b**) nighttime LST (°C), as well as annual mean NDVI values

daytime values (rho = 0.62, p-value = 0.00) and a positive but weak relationship with nighttime values (0.16, p-value = 0.49), although nonsignificant.

ECOSTRESS. ECOSTRESS-derived summer-mean 2018–2022 LSTs (Fig. 12.4) for the entire domain range from 9.69 to 55.37°C, with a mean of 27.11°C and a median of 27.42°C. The 0.5th percentile has a value of 19.63°C, whereas the 99.5th has a value of 29.6°C. The distributions of minimum values are found along the southwest towards the mountain over low-density areas in Saratoga and Los Gatos, while maximum LSTs are found east of San Jose over the Reid-Hillview Airport area. Lower than average values are located west towards Palo Alto and Los Altos Hills, where most of the developed open-space area is located. Sunnyvale and Santa Clara mostly exhibit average LST values over medium-to-high-intensity developed areas corresponding to commercial areas. Major roads, the San Jose Airport, and the southern part of the City of San Jose have higher-than-average summer LST values.

Summer-mean 2018–2022 ET values for the entire domain range from 6.14 to 418.01 Watts/m^2, with a mean of 129.73 Watts/m^2 and a median of 122.28 Watts/m^2. The 0.5th percentile has a value of 64.45 Watts/m^2, whereas the 99.5th has a value

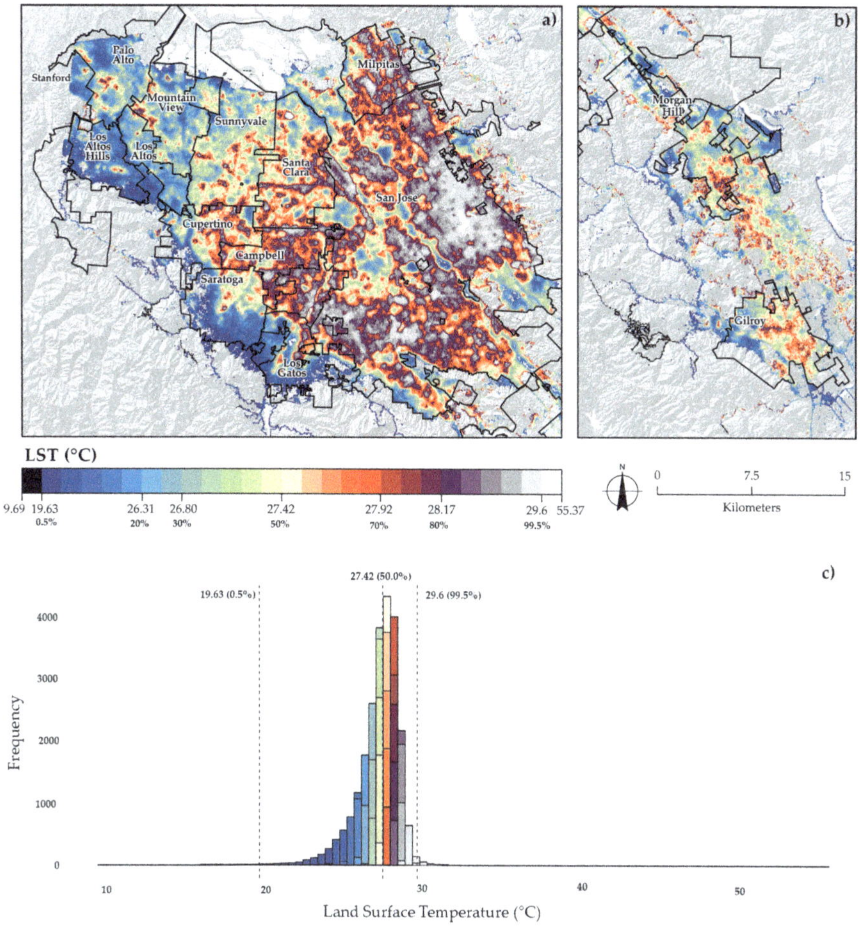

Fig. 12.4 ECOSTRESS (2018–2022) mean-summer (June–August) Land Surface Temperature (LST, °C) spatial distributions at 70-meter resolution for Santa Clara County in its (**a**) central and (**b**) southernmost parts, as well as (**c**) the frequency distribution of values

of 279.03 Watts/m^2. The spatial distributions (Fig. 12.5) show minimum values over the highly impervious Airport and Downtown areas in San Jose and to the southernmost region (6b) in Morgan Hill. Additional minimum ET areas are west in Cupertino, north in Milpitas, and the southernmost part of the study area in Gilroy over high-density and highly impervious areas. Maximum ET values are to the west in Palo Alto and south of Downtown San Jose over residential areas. Higher-than-average ET values are found in low-density and highly vegetated Los Altos, Saratoga, and Los Gatos. Lower-than-average ET values are primarily located east and south of San Jose over highly-vegetated residential areas.

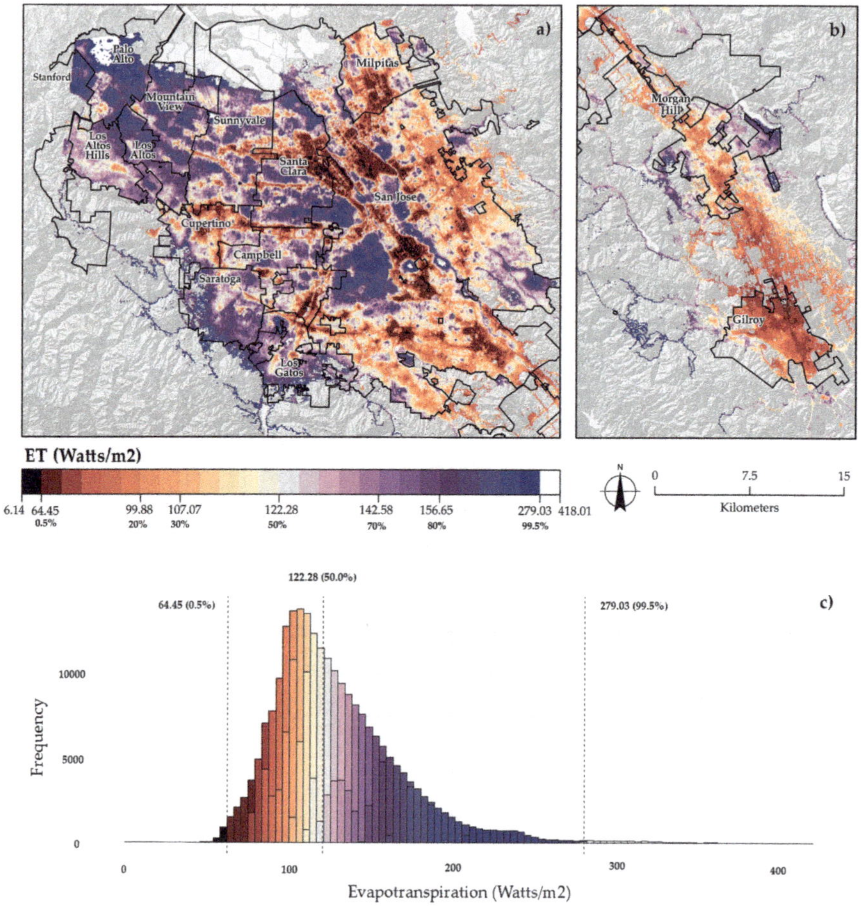

Fig. 12.5 ECOSTRESS (2018–2022) mean-summer (June–August) evapotranspiration (ET, Watts/m^2) spatial distributions at 70-meter resolution for Santa Clara County in its (**a**) central and (**b**) southernmost part. as well as (**c**) the frequency distribution of values

ET and LST results exhibit values that could be considered out of range (e.g., below the 0.5th and above 99.5th percentile thresholds). The resulting range of values is probably due to the cloud filtering process, where cloud pixels that were not recognized as clouds were not masked. An additional process to remove these undetected cloud pixels includes filtering outliers (e.g., eliminating pixels below the 8% or above 98% percentile). This process, however, was not done in the current effort to avoid over-filtering pixels. ET and LST values, however, exhibit similar spatial patterns, which thermal inertia can explain; surfaces with low ET values warm faster than their vegetated and well-watered surroundings. The Spearman correlation statistic, therefore, shows a strong negative and significant relationship between ET and LST values (rho = −0.62, p-value = 0.00).

12.3.2 Socioeconomic Environment

Local Climate Zones. The resulting LCZ classification yielded an overall accuracy of 0.73. The 100-meter LCZ distribution (Fig. 12.6) shows that low-rise urban structures (LCZs 3, 6, 8) comprise most of the urbanized area of Santa Clara County. LCZ 3 (compact low-rise) is primarily located within San Jose to its eastern and southernmost parts over highly-density and low-vegetated residential areas, where high LST values are also detected. LCZ 6 (open low-rise), in contrast, is mainly east of the study area over lower-density and highly-vegetated residential areas, while LCZ 8 (large low-rise) is found in the upper part of the study area towards the San Francisco Bay, where most commercial areas are located. LCZ 9 (sparsely built) surrounds the study area and is particularly prevalent in the southern portion of the County (Fig. 12.6b) around Morgan Hill, where most agricultural land is located. LCZ A (dense trees), B (scattered trees), and D (low plants) comprised most of the non-urban classes and are primarily found outside city limits.

Results of the evaluation between LSTs and non-urban classes (Fig. 12.7) show LCZ A (dense trees) have the lowest mean value of 22.2°C, while the maximum (27.9°C) is found in LCZ E (bare rock or paved). In urban LCZ classes, the minimum value (26.4°C) is found in LCZ 9 (sparsely built) and the maximum (28.1°C) in LCZ 3 (compact low-rise), which also exhibits minimum ranges (25.0 to 29.4°C). Maximum ranges (11.2 to 32.3°C) are found for LCZ 9 (sparsely built) due to their

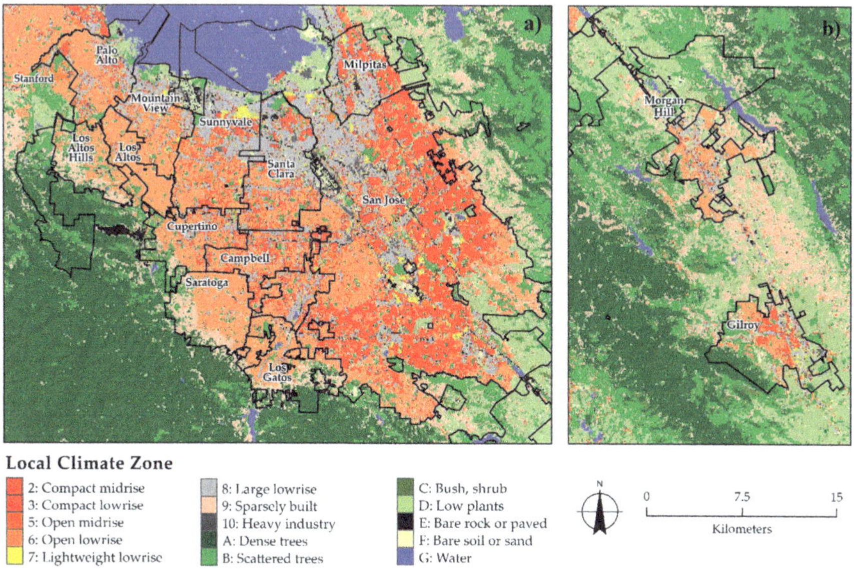

Fig. 12.6 Local Climate Zones distribution at a 100-m resolution for Santa Clara County in its (**a**) central and (**b**) southernmost parts

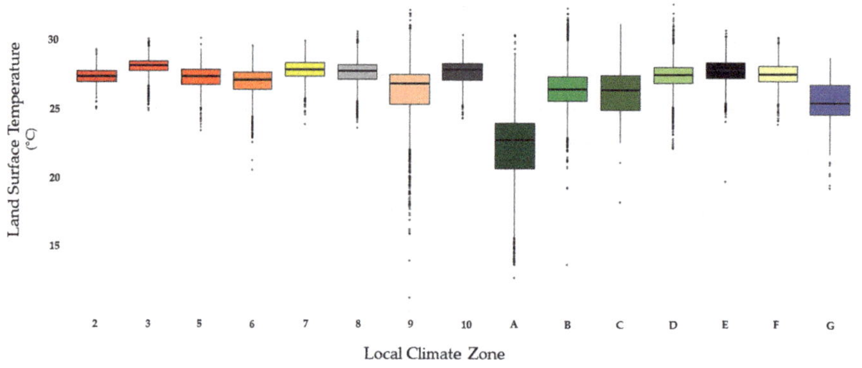

Fig. 12.7 Box plot of Land Surface Temperature (LST, °C) values per Local Climate Zone (LCZ) in Santa Clara County. The box represents the first (Q1) to third (Q3) quartile, and the horizontal line represents the median. Whiskers (vertical line) expand from minimum to maximum. Outliers (dots) are greater than Q3 + 1.5 IQR or less than Q1 - 1.5 IQR

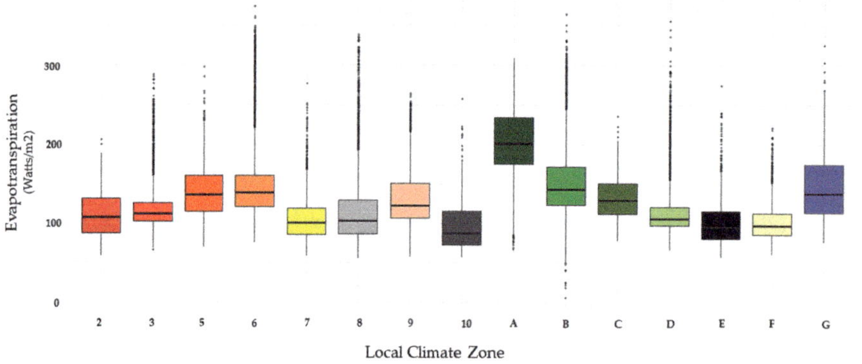

Fig. 12.8 Box plot of evapotranspiration (ET, Watts/m^2) values per Local Climate Zone (LCZ) in Santa Clara County. The box represents the first (Q1) to third (Q3) quartile, and the horizontal line represents the median. Whiskers (vertical line) expand from minimum to maximum. Outliers (dots) are greater than Q3 + 1.5 IQR or less than Q1 - 1.5 IQR)

spatial variability, reflecting the differences in the surrounding vegetation and elevation.

Results of the relationships between ET and non-urban LCZs (Fig. 12.8) show that the LCZ A (dense trees) class has the highest mean value (200 Watts/m^2) and LCZ E (bare rock or paved) has the lowest mean value (101 Watts/m^2). In urban areas, LCZ 6 (open low-rise) has the highest mean LST (144 Watts/m^2), while LCZ 10 (heavy industry) has the lowest (97 Watts/m^2). LZC 6 exhibits the largest range of values (75.5 to 374 Watts/m^2), whereas LCZ 2 (compact mid-rise) shows the lowest (59 to 205 Watts/m^2).

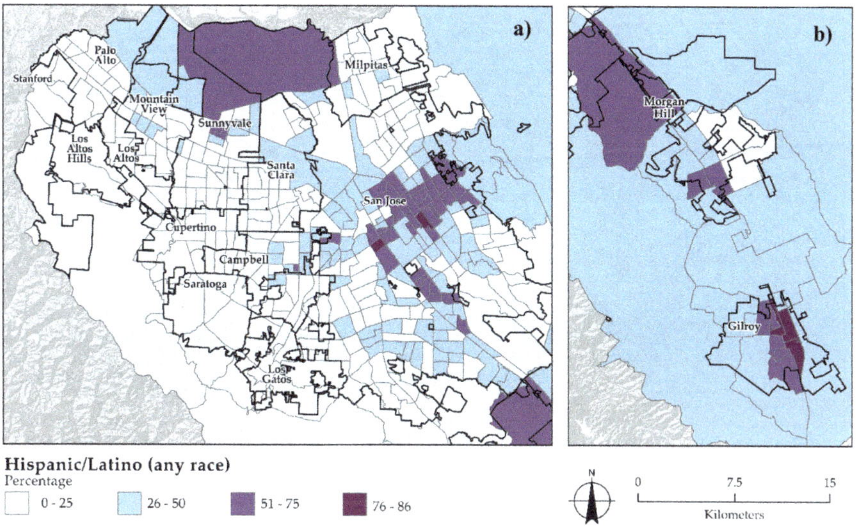

Fig. 12.9 Spatial distribution of Hispanic/Latino (any race) population as a percentage per census tract in Santa Clara County derived from US Census Bureau American Community Survey, Five-Year Estimates (2017–2021) data (US Census Bureau 2022)

Neighborhood demographics. In Santa Clara County, the percentage of Hispanic/Latinos per census tract varies from 0.38 to 85.6%, with a mean of 25.5%. Resulting distributions (Fig. 12.9) show a maximum concentration (76–86%) of Hispanic/Latino populations in two census tracts in San Jose and two in the southernmost region in Gilroy. These tracts are surrounded by higher-than-average (51–75%) concentrations of Hispanic/Latinos, with an additional census tract to the north in the below-sea level and residential Alviso neighborhood, where most salt marshes out to open waters of the San Francisco Bay are located. A slightly higher-than-average concentration (26–50%) is primarily found east of the study area in San Jose and to the south towards Morgan Hill and Gilroy. In the remaining tracts, mostly east of the County in Palo Alto, Los Altos, and Los Altos Hills, a lower-than-average (0–25%) concentration of Hispanic/Latino populations is found, as well as in the southwest in Los Gatos and Saratoga.

In Santa Clara County, median household income values range from $34,000 to more than $250,000, with an average value of $145,000. Median household income distributions (Fig. 12.10) show minimum values in the East San Jose neighborhoods and the southernmost region of the County in Gilroy, coinciding with areas with a maximum concentration of Hispanic/Latino populations. Maximum median household income values are found to the southwest in Los Gatos, southeast in Los Altos Hills, and east in Palo Alto. Higher-than-average median household income census tracts are mostly in Los Altos, Cupertino, and Mountain View, as well as in Campbell and the eastern part of the study area. Lower-than-average values are in the Downtown San Jose Area and some neighborhoods in the city of Santa Clara.

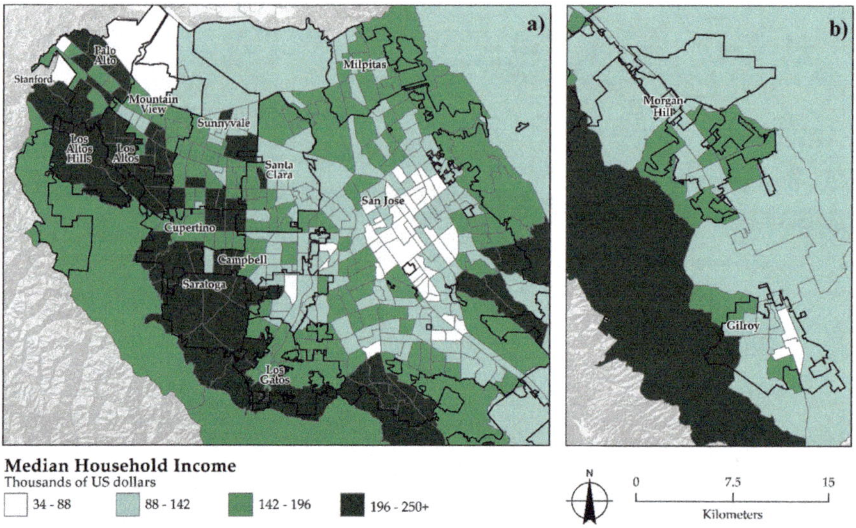

Fig. 12.10 Spatial distribution of median household income, in dollars, from the US Census Bureau American Community Survey, Five-Year Estimates (2017–2021) data (US Census Bureau 2022)

12.3.3 Human Environment

The Zonal Statistic Tool was used to calculate census-tract mean ET and LST to evaluate the human environment via the relationship of race/ethnicity and physical environment (ET and LST), with the use of a Spearman statistic. Results show a negative correlation between ET and LST (rho = −0.62, p-value = 0.00) and a strong negative correlation between the percentage of Hispanic/Latinos and median household income (rho = −0.78, p-value = 0.00). The percentage of Hispanic/Latinos negatively correlates with ET (rho = −0.35, p-value = 0.00) but positively correlates with LST (rho = 0.41, p-value = 0.00). A simple linear regression was performed to analyze rates of change of LST values by the percentage of the Hispanic/Latino population. Results show a positive LST-Hispanic/Latino trend (R^2 = 0.10), with LST values increasing by 0.2°C per percent increase in Hispanic/Latino.

The most frequently occurring LCZ was assigned to each census tract to analyze the relationship between sociodemographic variables and the built environment per the LCZ classification system. Results of this evaluation (Fig. 12.11) show that neighborhoods comprised of LCZ 2 (compact mid-rise) exhibit the lowest median household income mean ($115,000), followed by LCZ 3 (compact low-rise) with a median household income mean of $117,000. Maximum mean median household income values are in LCZ 9 (sparsely built) and LCZ A (dense trees), with $188,000 and $182,000, respectively. Neighborhoods comprised mainly of LCZ 3 (compact low-rise) and LCZ 8 (large low-rise) have the highest concentration of Hispanic/Latino populations (Fig. 12.11b). The lowest concentration of Hispanic/Latinos

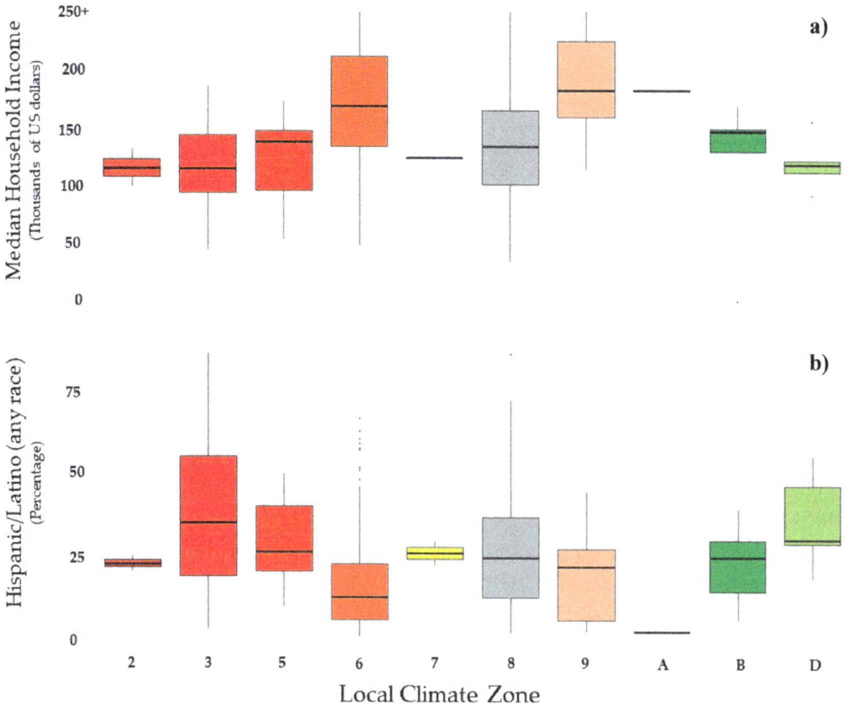

Fig. 12.11 Box plots of (**a**) Hispanic/Latino populations (percentage) and (**b**) median household income (in thousands of US dollars) per Local Climate Zone (LCZ) class in Santa Clara County. The box represents the first (Q1) to third (Q3) quartile, and the horizontal line represents the median. Whiskers (vertical line) expand from minimum to maximum. Outliers (dots) are greater than Q3 + 1.5 IQR or less than Q1 - 1.5 IQR

corresponds to LCZ A (dense trees) and LCZ 6 (sparsely built), which also exhibit the highest median household income values (Fig. 12.11a).

12.4 Discussion

Temporal analysis of LST in this study found results that were contrary to trend analysis reported previously for Santa Clara County. An analysis by Lebassi et al. (2009), from summer (June–August), long-term (1970–2005) daily 2-m air temperatures, reported a maximum trend of −0.53°C per decade for the study area. These trends resulted from marine air intrusions, as the air penetrated through a sea-level passage from the northern part of the San Francisco Bay southward towards Santa Clara County. As nighttime land surfaces cooled more rapidly than the sea, temperature differences produced maritime airflow. Kueppers et al. 2007 suggested that summer cooling in California resulted from increased irrigation, mainly from agricultural areas. Similarly, vegetation transpiring significant volumes of water within urban areas could provide higher cooling effects than trees with lower transpiration rates; thus, unirrigated trees or under water stress do not provide sufficient cooling (Ballinas and Barradas 2015).

In the urbanized area of the County, summer LSTs have an upward trend of 0.10°C per year (or 1.00°C per decade), whereas the corresponding nighttime values show no trend of 0.03°C per year (or 0.30 per decade). A higher proportion of the variance is explained for summer daytime (R^2 = 52%) than for corresponding nighttime values daytime (R^2 = 14%). The difference in summer trends between the current effort and those of Lebassi et al. (2009) may be attributed to measurement instruments. While this study uses remotely sensed data to measure LSTs, the latter uses National Climatic Data Center (NCDC) observational sites to measure daily T_{min} and T_{max} 2-m air temperatures to produce average monthly values.

This difference in trends may also be explained by Simpson's Paradox, which arises because aggregated data supports a trend that disappears or reverses when subgroups are combined (Pearl 2022). Lebassi et al. (2009) found that from 1970 to 2005, daily mean summer 2-m air temperatures increased by 0.15°C per decade. Observations were then aggregated into inland and coastal subgroups. The former showed an increasing trend of 0.24°C per decade, whereas the latter showed a decreasing trend of −0.01°C per decade. The aggregation of spatial data may also result in the Modifiable Areal Unit Problem, which arises when data is aggregated according to different zonal systems, causing inconsistent results.

During the summer season, NDVI values exhibit no trend, with a near-zero proportion of variance explained. During summer daytime, high NDVI values from plant growth provide maximum cooling via evapotranspiration, particularly in highly-irrigated areas, allowing them to cool faster. Vegetation also provides shade and absorbs radiation, cooling even over soil and impervious surfaces. A negative relationship consequently exists between summer LST and NDVI values during daytime and nighttime hours. During winter months, there is, however, a positive LST-NDVI relationship, and it is particularly stronger during daytime hours. The low photosynthesizing of vegetation provides no cooling and thus does not regulate LST values (Marzban et al. 2018).

During summer, irrigated vegetative cover in urban areas has high evapotranspiration values and thus warm at a slower rate than those with low evapotranspiration, hence explaining the negative relation. The spatial distribution of summer ET values shows high values in known high-income neighborhoods to the west in Palo Alto and Los Altos. Contrary results were found to the east in San Jose, where very low-income and majority Hispanic/Latino populations are located. Thermal inertia may also explain these differences in cooling rates: high-income neighborhoods tend to have high vegetative cover and, thus, warm and cool at slower rates than their impervious counterparts. Trees within high-income communities provide shade, which attenuates radiant energy flows through absorption and reflection, further cooling the surface and near-surface temperature (Declet-Barreto et al. 2013).

In the US, there is substantial evidence that a high proportion of low-income communities, where most racial and ethnic minorities reside, are deprived of green areas that resulted from historical ethno-racial oppression and segregated park system (Wolch et al. 2014). The percentage of Hispanic/Latino populations is higher in LCZ 3, which also exhibits the highest mean LST (28.1 °C). This study, however, does not disaggregate Hispanic/Latinos by origin, although there are socioeconomic differences among Hispanics, which may expose them to different levels of environmental hazards. Williams et al. (2010) report that Cubans have a college graduation rate three times higher than Mexicans, and Puerto Ricans have higher high school graduation rates than Mexicans (74 vs. 54%), which may contribute to health disparities and risk exposure. Hispanic/Latinos in the United States have been treated as a homogenous group, although differences have been reported within this population (Rivera et al. 2022). Mexicans and Cubans, for example, are less likely to have emergency department visits than non-Hispanic Whites (Weinick et al. 2004), potentially affecting their representation in rates of heat-related emergency room visits. Median household income operates with other characteristics like race, ethnicity, age, and gender. Additional sociodemographic variables, such as employment status, professional occupation, median house value, education levels, or occupational prestige, should be considered to for the analysis of human-environment relationships. Heat vulnerability studies have documented that minorities are overexposed to environmental hazards, including extreme heat. In California, for example, communities of color have lower median household income values, and fewer are homeowners (Morello-Frosch et al. 2002). They also have limited access to resources, support, information regarding heat protection actions, and mobility that could reduce their vulnerability (Wolf and McGregor 2013).

There is, however, a lack of consistency among variables used to measure heat exposure (climatic and built-up environment) and vulnerability (health, demographics, and socioeconomic status). A heat vulnerability index that considers the nuances of an area (e.g., ethnic/racial differences by origin), including its historical and institutional context (e.g., racial segregation), is required to identify individuals susceptible to extreme heat and that would facilitate place-to-place comparisons. The political and economic forces that lead to that environmental inequality must be examined to understand how institutional discrimination interacts with larger structural forces (Morello-Frosch et al. 2002). Studies that attempt to explain the geographical distribution of heat risk must emphasize the historical roots that caused today's health disparities among communities of color (Ford and Airhihenbuwa 2010).

There are also different levels of exposure to environmental racism. Thus, the level of heat exposure and perception will vary, to a certain extent, by age, gender, socioeconomic status among races (high-income Hispanic/Latino vs. high-income White), within races (e.g., high-income South American vs. low-income Mexican), immigration status, length of time in the US, health care access, and geographical location (Hispanic/Latino in California vs. Hispanic/Latino in Vermont). Intersectionality is a theoretical framework suggesting multiple social categories intersect at the individual level (Bowleg 2012), which may affect their heat vulnerability. The intersection of race and income was studied in the context of fluid intake of individuals (Brooks et al. 2017). Results showed Hispanic/Latino heritage and non-Hispanic Black people in the United States are higher than for non-Hispanic

Whites, while lower-income adults had a higher risk of inadequate hydration than higher-income adults. Heat studies that consider how differences in the built environment, individual bodies (e.g., children, elderly, people with mobility constraints, pregnant women), and social characteristics (e.g., race, gender, socioeconomic status) intersect are necessary to understand how these moderate heat experiences, including physiologically and behavioral responses (Hamstead 2023).

12.5 Conclusion

As urbanization accelerates, the built environment expands and replaces natural land cover with human-made surfaces and heat-emanating activities. This ongoing process will continue exacerbating the formation of UHIs, with harmful health effects on local populations, particularly the vulnerable. The spatial identification of these at-risk areas is paramount for implementing effective mitigation and adaptation actions to reduce heat stress. The study of population exposure, however, involves multi-temporal and multi-spatial human and environmental processes and, thus, renders some sensors insufficient to understand these human-environment interactions.

The present study suggests an exploratory methodology to explore the potential association of LST-NDVI/ET values and race/ethnicity- median household income to evaluate the human-environment spatial relationships between heat exposure and evaporative cooling for each group, as well as differences in the built environment. The physical environment was assessed from a temporal and spatial perspective via the acquisition of MODIS and ECOSTRESS. The built environment was evaluated with the use of the LCZ framework. The socioeconomic environment was assessed with the use of the median household income and Hispanic/Latino (any race) variables acquired from the US American Community Survey 2017–2021 census data.

Significant results from this effort show that for the urbanized area of Santa Clara County, a positive summer-daytime LST trend of 0.10°C per year but a weak LST trend for corresponding nighttime hours (0.03°C per year). A negative relation between summer LST and NDVI values was found, reflecting the cooling effect of vegetation. During winter, a positive LST-NDVI relationship was found, particularly for daytime hours, given that the low photosynthesizing of vegetation provides no cooling, thus not regulating LST values. A strong negative relationship was also found for ECOSTRESS-derived summer LST and ET values. The spatial distribution shows minimum LSTs and maximum ET values in high-income communities like Los Altos Hills, Los Altos, and Palo Alto, where a low percentage of Hispanic/Latino populations is found. Maximum ET values and minimum LSTs primarily occurred in low-income neighborhoods of San Jose, with a maximum concentration of Hispanic/Latinos. Low median household income populations ($117,000) and Hispanic/Latinos most likely exist within LCZ 3, which exhibits the highest mean LST (28.1°C).

Future efforts should overcome some of the limiting assumptions required for the current effort, e.g., the use of monthly mean LSTs instead of hourly 2-m observations. While the latter are consistent with local government heat warnings, they only provide limited spatial coverage. A technique thus must be developed to combine the 2-m values, satellite-derived LST distributions, and fine-scale 2-m values from Weather Research and Forecasting model simulations. Surrounding vegetation differences within LCZ classes must also be considered to establish LCZ-LST relationships better. Future multidisciplinary efforts should also consider an intersectionality framework for the study of heat vulnerability, as well as differences between and within racial/ethnic and socioeconomic groups, to predict future heat risks under various climate change scenarios. The identification of areas at high risk of heat stress would facilitate the targeting of public health programs and other mitigation strategies, such as the increase of urban vegetation or the establishment of cooling centers (Widerynski et al. 2017), which could provide a cool environment for vulnerable individuals in Santa Clara County and elsewhere experiencing similar intra-urban human-environment phenomena.

12.6 Design Implications

The reduction of urban evaporative cooling is an important factor contributing to surface UHIs. Urban moisture content is lower than in surrounding areas, allowing them to warm faster than their vegetated counterparts. The results of this study provide a methodology for local governments to evaluate high-temperature distributions and implement mitigation/adaptation strategies to reduce heat risks (urban greenery projects, zoning codes, and air quality standards), particularly in low-income communities whose residents are most vulnerable.

Heat mitigation actions are important, but they are not the definite solution if unequal heat exposure continues disadvantaging groups based on income, race, or color. Government agencies must identify areas where human health and the environment are disproportionately affected by heat and other noxious facilities. Programs, policies, and zoning ordinances must then be placed to redress inequities and allocate resources and actions where the environmental and health needs are the greatest. Even if green areas are required by building and construction codes, evidence suggests that zoning ordinance enforcement is low within low-income communities, especially in areas where homeownership is low.

Although green redevelopment can mitigate the increasing urban heat effects, it can also exclude those whose need for access is most acute. As access to green space is recognized as environmental justice, cities have implemented the supply of urban green areas in low-income areas, which has made them attractive to a population of higher status, encouraging gentrification. Access to credits for landlords to upgrade housing materials in low-income communities could decrease the negative impacts of high summer temperatures without promoting green gentrification.

References

Allen MA, Roberts DA, McFadden JP (2021) Reduced urban green cover and daytime cooling capacity during the 2012–2016 California drought. Urban Clim *36*(December 2020):100768. https://doi.org/10.1016/j.uclim.2020.100768

Ballinas M, Barradas VL (2015) The urban tree as a tool to mitigate the urban heat Island in Mexico city: a simple phenomenological model. J Environ Qual 45(1):157. https://doi.org/10.2134/jeq2015.01.0056

Bechtel B, Alexander P, Böhner J, Ching J, Conrad O, Feddema J, Mills G, See L, Stewart I (2015) Mapping local climate zones for a worldwide database of the form and function of cities. ISPRS Int J Geo-Inf 4(1):199–219. https://doi.org/10.3390/ijgi4010199

Bornstein RD (1968) Observations of the urban heat island effect in New York City. J Appl Meteorol 7(4):575–582. https://doi.org/10.1175/1520-0450(1968)007<0575:OOTUHI>2.0.CO;2

Bornstein R, Styrbicki-Imamura R, González JE, Lebassi B (2012) Interactions of global-warming and urban heat islands in different climate-zones. In: Fernando HJS, Klaić Z, McCulley JL (eds) National security and human health implications of climate change. Springer, Dordrecht, pp 49–60

Bowleg L (2012) The problem with the phrase women and minorities: intersectionality-an important theoretical framework for public health. Am J Public Health 102(7):1267–1273. https://doi.org/10.2105/AJPH.2012.300750

Brooks CJ, Gortmaker SL, Long MW, Cradock AL, Kenney EL (2017) Racial/ethnic and socioeconomic disparities in hydration status among US adults and the role of tap water and other beverage intake. Am J Public Health 107(9):1387–1394. https://doi.org/10.2105/AJPH.2017.303923

Budhiraja B, Agrawal G, Pathak P (2020) Urban heat island effect of a polynuclear megacity Delhi – compactness and thermal evaluation of four sub-cities. Urban Clim *32*(May 2019):100634. https://doi.org/10.1016/j.uclim.2020.100634

Cai M, Ren C, Xu Y, Lau K, Wang R (2018) Investigating the relationship between local climate zone and land surface temperature using an improved WUDAPT methodology – a case study of Yangtze River Delta, China. Urban Clim *24*:485–502. https://doi.org/10.1016/j.uclim.2017.05.010

Cawse-Nicholson K, Anderson MC, Yang Y, Yang Y, Hook SJ, Fisher JB, Halverson G, Hulley GC, Hain C, Baldocchi DD, Brunsell NA, Desai AR, Griffis TJ, Novick KA (2021) Evaluation of a CONUS-Wide ECOSTRESS DisALEXI evapotranspiration product. IEEE J Sel Top Appl Earth Obs Remote Sens 14:10117–10133. https://doi.org/10.1109/JSTARS.2021.3111867

Chakraborty T, Lee X (2019) A simplified urban-extent algorithm to characterize surface urban heat islands on a global scale and examine vegetation control on their spatiotemporal variability. Int J Appl Earth Obs Geoinf 74(May 2018):269–280. https://doi.org/10.1016/j.jag.2018.09.015

Chakraborty T, Hsu A, Manya D, Sheriff G (2020) A spatially explicit surface urban heat island database for the United States: characterization, uncertainties, and possible applications. ISPRS J Photogramm Remote Sens 168(April):74–88. https://doi.org/10.1016/j.isprsjprs.2020.07.021

Chang Y, Xiao J, Li X, Zhou D, Wu Y (2022) Combining GOES-R and ECOSTRESS land surface temperature data to investigate diurnal variations of surface urban heat island. Sci Total Environ 823:153652. https://doi.org/10.1016/j.scitotenv.2022.153652

Ching J, Mills G, Bechtel B, See L, Feddema J, Wang X, Ren C, Brorousse O, Martilli A, Neophytou M, Mouzourides P, Stewart I, Hanna A, Ng E, Foley M, Alexander P, Aliaga D, Niyogi D, Shreevastava A et al (2018) WUDAPT: an urban weather, climate, and environmental modeling infrastructure for the anthropocene. Bull Am Meteorol Soc 99(9):1907–1924. https://doi.org/10.1175/BAMS-D-16-0236.1

Coleman RW, Stavros N, Hulley G, Parazoo N (2020) Comparison of thermal infrared-derived maps of irrigated and non-irrigated vegetation in urban and non-urban areas of southern California. Remote Sens 12(24):1–19. https://doi.org/10.3390/rs12244102

Declet-Barreto J, Brazel AJ, Martin CA, Chow WTL, Harlan SL (2013) Creating the park cool island in an inner-city neighborhood: heat mitigation strategy for Phoenix, AZ. Urban Ecosyst 16(3):617–635. https://doi.org/10.1007/s11252-012-0278-8

Demuzere M, Kittner J, Bechtel B (2021) LCZ generator: a web application to create local climate zone maps. Front Environ Sci 9:637455. https://doi.org/10.3389/fenvs.2021.637455

Dong L, Mitra C, Greer S, Burt E (2018) The dynamical linkage of atmospheric blocking to drought, heatwave and urban heat island in southeastern US: a multi-scale case study. Atmosphere 9(1). https://doi.org/10.3390/ATMOS9010033

Dou J, Wang Y, Bornstein R, Miao S (2015) Observed spatial characteristics of Beijing urban climate impacts on summer thunderstorms. J Appl Meteorol Climatol 54(1):94–105. https://doi.org/10.1175/JAMC-D-13-0355.1

Duckworth FS, Sandberg JS (1954) The effect of cities upon horizontal and vertical temperature gradients. Bull Am Meteorol Soc 35(5):198–207. http://www.jstor.org/stable/26242101

Ermida SL, Soares P, Mantas V, Göttsche FM, Trigo IF (2020) Google earth engine open-source code for land surface temperature estimation from the landsat series. Remote Sens 12(9):1–21. https://doi.org/10.3390/RS12091471

Ford CL, Airhihenbuwa CO (2010) Critical race theory, race equity, and public health: toward antiracism praxis. Am J Public Health 100(S1):S30–S35. https://doi.org/10.2105/AJPH.2009.171058

Gorelick N, Hancher M, Dixon M, Ilyushchenko S, Thau D, Moore R (2017) Google Earth Engine: planetary-scale geospatial analysis for everyone. Remote Sens Environ 202:18–27. https://doi.org/10.1016/j.rse.2017.06.031

Hamstead ZA (2023) Critical heat studies: deconstructing heat studies for climate justice. Plan Theory Pract 24(2):153–172. https://doi.org/10.1080/14649357.2023.2201604

He BJ (2018) Potentials of meteorological characteristics and synoptic conditions to mitigate urban heat island effects. Urban Clim *24*(November 2017):26–33. https://doi.org/10.1016/j.uclim.2018.01.004

Imamura IR (1991) Observational studies of urban heat Island characteristics in different climate zones. Ph.D. thesis. University of Tsukuba, p 145

Imhoff ML, Zhang P, Wolfe RE, Bounoua L (2010) Remote sensing of the urban heat island effect across biomes in the continental USA. Remote Sens Environ 114(3):504–513. https://doi.org/10.1016/j.rse.2009.10.008

IPCC (2014) Climate change 2014: synthesis report. Contribution of working groups I, II and III to the fifth assessment report of the intergovernmental panel on climate change [Core Writing Team, Pachauri RK, Meyer LA (eds)]. IPCC, Geneva, Switzerland, p 151

Jauregui E (1997) Heat island development in Mexico City. Atmos Environ *31*(22):3821–3831. https://doi.org/10.1016/S1352-2310(97)00136-2

Kueppers LM, Snyder MA, Sloan LC (2007) Irrigation cooling effect: regional climate forcing by land-use change. Geophys Res Lett *34*(3):1–5. https://doi.org/10.1029/2006GL028679

Lebassi B, Gonzalez J, Fabris D, Maurer E, Miller N, Milesi C, Switzer P, Bornstein R (2009) Observed 1970-2005 cooling of summer daytime temperatures in coastal California. J Clim 22(13):3558–3573. https://doi.org/10.1175/2008JCLI2111.1

Li L, Zha Y, Zhang J (2020) Spatial and dynamic perspectives on surface urban heat island and their relationships with vegetation activity in Beijing, China, based on Moderate Resolution Imaging Spectroradiometer data. Int J Remote Sens 41(3):882–896. https://doi.org/10.1080/01431161.2019.1650985

Manoli G, Fatichi S, Bou-Zeid E, Katul GG (2020) Seasonal hysteresis of surface urban heat islands. Proc Natl Acad Sci USA 117(13):7082–7089. https://doi.org/10.1073/pnas.1917554117

Marzban F, Sodoudi S, Preusker R (2018) The influence of land-cover type on the relationship between NDVI–LST and LST-T air. Int J Remote Sens 39(5):1377–1398

McMichael AJ (2000) The urban environment and health in a world of increasing globalization: issues for developing countries. Bull World Health Organ 78(9):1117–1126. https://doi.org/10.1590/S0042-96862000000900007

Miller DL, Wetherley EB, Roberts DA, Tague CL, McFadden JP (2022) Vegetation cover change during a multi-year drought in Los Angeles. Urban Clim 43(February):101157. https://doi.org/10.1016/j.uclim.2022.101157

Mini C, Hogue TS, Pincetl S (2014) Estimation of residential outdoor water use in Los Angeles, California. Landsc Urban Plan 127:124–135. https://doi.org/10.1016/j.landurbplan.2014.04.007

Morello-Frosch R, Pastor M, Porras C, Sadd J (2002) Environmental justice and regional inequality in southern California: implications for future research. Environ Health Perspect 110(2):149–154. https://doi.org/10.1289/ehp.02110s2149

Moseley WG, Perramond E, Hapke HM, Laris P (2013) An introduction to human-environment geography: local dynamics and global processes. Germany: Wiley

Ng SC, Low KS, Tioh NH (2011) Newspaper sandwiched aerated lightweight concrete wall panels - thermal inertia, transient thermal behavior and surface temperature prediction. Energ Buildings 43(7):1636–1645. https://doi.org/10.1016/j.enbuild.2011.03.007

Oke TR (1982) The energetic basis of the urban heat island. Q J Roy Meteorol Soc 108(455):1–24. https://doi.org/10.1002/qj.49710845502

Oke TR (1987) Boundary layer climates. Routledge. https://doi.org/10.4324/9780203407219

Oke T, Mills G, Christen A, Voogt J (2017) Urban climates. In: Urban climates. Cambridge University Press. https://doi.org/10.1017/9781139016476

Pearl J (2022) Comment: understanding Simpson's paradox. In: Probabilistic and causal inference: the works of Judea Pearl, pp 399–412

Peng S, Piao S, Ciais P, Friedlingstein P, Ottle C, Bréon FM, Nan H, Zhou L, Myneni RB (2012) Surface urban heat island across 419 global big cities. Environ Sci Technol 46(2):696–703. https://doi.org/10.1021/es2030438

Raj S, Paul SK, Chakraborty A, Kuttippurath J (2020) Anthropogenic forcing exacerbating the urban heat islands in India. J Environ Manage 257(August 2019):110006. https://doi.org/10.1016/j.jenvman.2019.110006

Ren C, Cai M, Li X, Zhang L, Wang R, Xu Y, Ng E (2019) Assessment of local climate zone classification maps of cities in China and feasible refinements. Sci Rep 9(1):1–11. https://doi.org/10.1038/s41598-019-55444-9

Rivera E, Antonio-Némiga X, Origel-Gutiérrez G, Sarricolea P, Adame-Martínez S (2017) Spatiotemporal analysis of the atmospheric and surface urban heat islands of the Metropolitan Area of Toluca, Mexico. Environ Earth Sci 76(5):225. https://doi.org/10.1007/s12665-017-6538-4

Rivera A, Darden JT, Dear N, Grady SC (2022) Environmental injustice among Hispanics in Santa Clara, California: a human–environment heat vulnerability assessment. GeoJournal. https://doi.org/10.1007/s10708-022-10768-4

Siddiqui A, Kushwaha G, Nikam B, Srivastav SK, Shelar A, Kumar P (2021) Analysing the day/night seasonal and annual changes and trends in land surface temperature and surface urban heat island intensity (SUHII) for Indian cities. Sustain Cities Soc 75(August):103374. https://doi.org/10.1016/j.scs.2021.103374

Sismanidis P, Bechtel B, Perry M, Ghent D (2022) The seasonality of surface urban heat islands across climates. Remote Sens 14(10):1–21. https://doi.org/10.3390/rs14102318

Stewart ID, Oke TR (2012) Local climate zones for urban temperature studies. Bull Am Meteorol Soc *93*(12):1879–1900. https://doi.org/10.1175/BAMS-D-11-00019.1

Stewart ID, Mills G (2021) The urban heat Island. Elsevier

Swaid H (1991) Nocturnal variation of air-surface temperature gradients for typical urban and rural surfaces. Atmos Environ Part B *25*(3):333–341. https://doi.org/10.1016/0957-1272(91)90005-Y

Taha H (2017) Characterization of urban heat and exacerbation: development of a heat island index for California. Climate *5*(3):59. https://doi.org/10.3390/cli5030059

Tayyebi A, Darrel Jenerette G (2016) Increases in the climate change adaption effectiveness and availability of vegetation across a coastal to desert climate gradient in metropolitan Los Angeles, CA, USA. Sci Total Environ 548–549:60–71. https://doi.org/10.1016/j.scitotenv.2016.01.049

Thanabalan P, Vidhya R, Kankara RS, Manonmani R (2023) Time-series analysis of MODIS (LST and NDVI) and TRMM rainfall for drought assessment over India. Appl Geomatics 383–405. https://doi.org/10.1007/s12518-023-00505-y
US Census Bureau (2022) 2017–2021 American Community Survey, 5-Year Estimates. Table, DP03 and DP05; generated by Ana Rivera; https://data.census.gov/ (February 26, 2022)
USGS (2019) NLCD 2011 land cover. National Geospatial Data Asset (NGDA) land use land cover. U.S. Geological Survey. Retrieved from https://www.mrlc.gov/data
Verdonck ML, Demuzere M, Hooyberghs H, Beck C, Cyrys J, Schneider A, Dewulf R, Van Coillie F (2018) The potential of local climate zones maps as a heat stress assessment tool, supported by simulated air temperature data. Landscape Urban Plann *178*(July 2017):183–197. https://doi.org/10.1016/j.landurbplan.2018.06.004
Voogt J (2007) How researchers measure urban heat islands. United States Environmental Protection Agency (EPA), State and local climate and energy program, heat island effect, urban heat island webcasts and conference calls, 34. http://epa.gov/heatisland/resources/pdf/EPA_How_to_measure_a_UHI.pdf
Wang M, Zhang Z, Hu T, Wang G, He G, Zhang Z, Li H, Wu Z, Liu X (2020) An efficient framework for producing landsat-based land surface temperature data using google earth engine. IEEE J Sel Top Appl Earth Obs Remote Sens *13*:4689–4701. https://doi.org/10.1109/JSTARS.2020.3014586
Wang C, Ren Z, Chang X, Wang G, Hong X, Dong Y, Guo Y, Zhang P, Ma Z, Wang W (2023) Understanding the cooling capacity and its potential drivers in urban forests at the single tree and cluster scales. Sustain Cities Soc 93(March):104531. https://doi.org/10.1016/j.scs.2023.104531
Weinick RM, Jacobs EA, Stone LC, Ortega AN, Burstin H (2004) Hispanic healthcare disparities challenging the myth of a monolithic hispanic population. Med Care 42(4):313–320. https://doi.org/10.1097/01.mlr.0000118705.27241.7c
Widerynski S, Schramm PJ, Conlon KC, Noe RS, Grossman E, Hawkins M, Nayak SU, Roach M, Hilts AS (2017) Use of cooling centers to prevent heat-related illness: summary of evidence and strategies for implementation. https://stacks.cdc.gov/view/cdc/47657
Williams DR, Mohammed SA, Leavell J, Collins C (2010) Race, socioeconomic status, and health: complexities, ongoing challenges, and research opportunities. Ann N Y Acad Sci 1186(1):69–101. https://doi.org/10.1111/j.1749-6632.2009.05339.x
Wolch JR, Byrne J, Newell JP (2014) Urban green space, public health, and environmental justice: the challenge of making cities "just green enough.". Landsc Urban Plan *125*:234–244. https://doi.org/10.1016/j.landurbplan.2014.01.017
Wolf T, McGregor G (2013) The development of a heat wave vulnerability index for London, United Kingdom. Weather Clim Extremes 1(August 2003):59–68. https://doi.org/10.1016/j.wace.2013.07.004
Wu X, Wang G, Yao R, Wang L, Yu D, Gui X (2019) Investigating surface urban heat islands in South America based on MODIS data from 2003–2016. Remote Sens 11(10). https://doi.org/10.3390/rs11101212

Chapter 13
Towards Drought-Sensitive Cities

Rob Roggema

Abstract In increasingly dryer and hotter conditions cities need to adapt in the (near) future. To design drought-sensitive cities much can be learned from the program of water-sensitive cities. In this program a research center was established, policy-implications have been derived and the findings were applied in many local projects. A drought-sensitive program should link research, governance, and practice. Moreover, to fully integrate drought conditions in the development of our cities, the future dryer conditions should be anticipated, embraced and, if possible reversed. This needs to be done by connecting long-term visions with short-term interventions and to integrate spatial projects at the regional and larger scale with hyperlocal spatial transformations, which are visible for urban residents and make a difference for their daily livability.

Keywords Drought-sensitive city · Embracing change · Anticipating future

In a rapidly warming world, cities suffer the most, due to their large portion of concrete, asphalt, bricks, glass facades and impermeable surfaces. The higher temperatures not only bring heat but also droughts. Without water a range of problems arise. In the thirsty city (Kumar et al. 2024) suffers from water scarcity and water stress, and biological processes slow down. Plants, animals, and humans have difficulties to live and survive in these conditions.

In urban design, planning, and development there is ample attention for the impacts of drought circumstances on the city and its public spaces, or the possibilities to design the urban environment in a way to create a drought sensitive city.

Water sensitive cities (Wong and Brown 2009; Rogers et al. 2020) is an established concept, with research centers (such as the inaugural one the CRC water sensitive cities in Melbourne), policies and an abundance of design projects dealing

R. Roggema (✉)
Escuela de Arquitectura, Artes y Diseño, Tecnológico de Monterrey, Monterrey, Nuevo León, Mexico
e-mail: rob@cittaideale.eu

R. Roggema (ed.), *Dry Urbanism*, Contemporary Urban Design Thinking,
https://doi.org/10.1007/978-3-031-77347-1_13

with excess of water. Plans to create a water sensitive Mexico City (Marín Salinas et al. 2016), local water sensitive design projects such as the water squares in Rotterdam (Ilgen et al. 2019; Elewa 2019), or river catchment approaches such as the Room for the River program (Klijn et al. 2013; Zevenbergen et al. 2015).

Where water sensitive cities are well-understood and in operation, the dry counterpart has not yet reached such a status. Can we create drought-sensitive cities, and if so, what can we learn from the water-sensitive cities approach?

First and foremost, it is not enough to only respond or cope with the conditions, whether they'd be wet or dry. Reacting to a drought of often too late to create a humid, healthy, and livable environment. Therefore, it is needed to anticipate future conditions. The knowledge that it can get dryer and hotter in the (near) future should be taken serious and lead to alternative and novel approaches. This means integrating the opportunities to catch water, create humid conditions and enhancing shade in urban development.

Beyond anticipation, the drought can also be embraced. Instead of seeing it as a threat, the key elements, both environmental as cultural can be used to integrate in the urban design. The specific cultural and traditional understandings of dry conditions can be implemented in current cities through xeriscapes that emphasize the natural dry and hot landscapes and give these the space in the city. This way local knowledge and local native species can flourish and provide an environment that is adaptive to changing climates and gives the neighborhoods an ecology that fits with the heritage. Moreover, the cultural dimension of local traditions can be integrated in design options to stay close to the benefits of the local knowledge and urban and architectural principles.

Anticipation and embracing focus on dealing with dry conditions. Alternatively, we can implement interventions that could reduce heating and droughts. Large scale forests, which can autonomously grow through a biotic pump emergence (Makarieva and Gorshkov 2006, 2007) could reduce regional temperatures and decrease local temperature rise.

For the design of cities, this means that we must work at different time and spatial scales. It doesn't make sense to have a long-term strategy without any short-term actions. Also, it doesn't make sense to have a large-scale vision without any implications locally. But the opposite is also true, many local and short-term initiatives do not create the desired change without a vision on the long term and larger scale.

So, what do we need to do to create cities that can live with dry and hot conditions?

1. Connect the large-scale and long-term with the local scale and short-term.
2. Integrate research programs, policy documents with real actionable projects. This implies a collaboration between academia, the government, enterprises, and the residents.
3. Design the city to anticipate droughts, embrace and reverse them.

How would a drought-sensitive city look like? First, it combats droughts by creating places where humidity is enhanced. Wetlands, forests, and trees enhance humidity. Secondly, the city should create space for xeriscapes, where dry habitats can flourish. These spaces should be generous and abundant, to enrich the city with valuable ecosystems. Thirdly, the city should incorporate large scale projects to reverse dry and hot conditions.

The city of drought demands an alternative approach. We are at the forefront of novel thinking on how to accommodate and design cities that face increased problems caused by higher temperatures and droughts. This also means that urban designers and landscape architects must accommodate this quest by proposing designs that strategize and localize solutions to improve local livability for people, non-human and inanimal creatures.

References

Elewa AKA (2019) Flexible public spaces through spatial urban interventions, towards resilient cities. Eur J Sustain Dev 8(4):152–152

Ilgen S, Sengers F, Wardekker A (2019) City-to-city learning for urban resilience: the case of water squares in Rotterdam and Mexico City. Water 11(5):983

Klijn F, de Bruin D, de Hoog MC, Jansen S, Sijmons DF (2013) Design quality of room-for-the-river measures in The Netherlands: role and assessment of the quality team (Q-team). Int J River Basin Manag 11(3):287–299

Kumar M, Roggema R, Dogra S, Aguilar-Barajas I, Snyder SA, Furumai H, Mahlknecht J (2024) Global rise of thirsty cities: defining and redesigning the strategies for water sustainability. ACS EST Water 4:2775–2778. https://doi.org/10.1021/acsestwater.4c00362

Makarieva AM, Gorshkov VG (2006) Biotic pump of atmospheric moisture as driver of the hydrological cycle on land. Hydrol Earth Syst Sci Discuss 3(4):2621–2673

Makarieva AM, Gorshkov VG (2007) Biotic pump of atmospheric moisture as driver of the hydrological cycle on land. Hydrol Earth Syst Sci 11(2):1013–1033

Marín Salinas E, Boer F, Van de Pas B, Rico Espínola V (2016) Towards a watersensitive Mexico City. Public space as a rain management strategy. AEP, De Urbanisten, Deltares, Rotterdam. Available online: https://www.deltares.nl/app/uploads/2018/01/20160629_WS-CDMX_final-version-report-smsize.pdf. Accessed 5 Sept 2024

Rogers BC, Dunn G, Hammer K, Novalia W, De Haan FJ, Brown L, Brown RR, Lloyd S, Urich C, Wong THF, Chesterfield C (2020) Water sensitive cities index: a diagnostic tool to assess water sensitivity and guide management actions. Water Res 186:116411. https://doi.org/10.1016/j.watres.2020.116411

Wong TH, Brown RR (2009) The water sensitive city: principles for practice. Water Sci Technol 60(3):673–682

Zevenbergen C, Rijke J, Van Herk S, Bloemen PJTM (2015) Room for the river: a stepping stone in adaptive delta management. Int J Water Gov 3(3):121–140

Index

R. Roggema (ed.), *Dry Urbanism*, Contemporary Urban Design Thinking,
https://doi.org/10.1007/978-3-031-77347-1

GPSR Compliance

The European Union's (EU) General Product Safety Regulation (GPSR) is a set of rules that requires consumer products to be safe and our obligations to ensure this.

If you have any concerns about our products, you can contact us on ProductSafety@springernature.com

In case Publisher is established outside the EU, the EU authorized representative is:

Springer Nature Customer Service Center GmbH
Europaplatz 3
69115 Heidelberg, Germany

Batch number: 07973899

Printed by Printforce, the Netherlands